South-South Collaboration in Health Biotechnology

South-South Collaboration in Health Biotechnology

Growing Partnerships amongst Developing Countries

Editor:
Halla Thorsteinsdóttir

ACADEMIC FOUNDATION
NEW DELHI

International Development Research Centre
Ottawa • Cairo • Dakar • Montevideo • Nairobi • New Delhi • Singapore

First published in 2012
by

ACADEMIC FOUNDATION
4772-73 / 23 Bharat Ram Road, (23 Ansari Road),
Darya Ganj, New Delhi - 110 002 (India).
Phones : 23245001 / 02 / 03 / 04.
Fax : +91-11-23245005.
E-mail : books@academicfoundation.com
www.academicfoundation.com

and

INTERNATIONAL DEVELOPMENT RESEARCH CENTRE
PO Box 8500,
Ottawa, ON K1G 3H9
Canada
E-mail: info@idrc.ca
www.idrc.ca

Cataloging in Publication Data--DK
Courtesy: D.K. Agencies (P) Ltd. <docinfo@dkagencies.com>

South-South collaboration in health biotechnology : growing partnerships amongst developing countries / editor, Halla Thorsteinsdóttir.
p. cm.
ISBN 13: 9788171889099
ISBN (e-book) 9781552505359

1. Pharmaceutical biotechnology--Developing countries. 2. Public health--Developing countries. 3. Cooperation--Developing countries. I. Thorsteinsdóttir, Halla. II. International Development Research Centre (Canada)

DDC 362.1091724 23

Typeset by Italics India, New Delhi.
Printed and bound in India.

Contents

Foreword

This book presents a detailed portrait of the contours of South-South collaboration in the health biotechnology sector. In particular, it casts revealing light on the factors that guide effective scientific partnerships and exchanges. The authors explore these issues by combining a wide range of quantitative and qualitative methodologies, including co-publications analyses, in-depth surveys of biotechnology firms and interviews with around 350 researchers, entrepreneurs and policy-makers in developing countries.

The key findings indicate that the level of South-South collaboration among researchers in health biotechnology remains low but is slowly increasing and that entrepreneurial collaboration seems to be more prevalent. Collaboration has helped to extend capacity in health biotechnology research, manufacturing and innovation to an increasing number of developing countries and thereby lessened the divide between them. Such collaboration has strongly focussed on shared health needs and has helped to increase the availability of more affordable health products and services. Governments and non-governmental organisations (NGOs) have also been able to foster closer ties between researchers by establishing programmes and extending funding for collaboration. Nevertheless there is still a lack of dedicated resources.

The authors call on governments in developed countries and international or philanthropic organisations to promote South-South collaboration as a means of enhancing development and encouraging global health. The authors also call on developing countries to take better advantage of this collaboration by more closely integrating these efforts with their own initiatives for scientific capacity building and improved health care.

This book makes valuable contributions to the wide-ranging scholarly literature on how to strengthen the impact of South-South collaboration for the benefit of people in need.

March 7, 2011 — Mohamed H.A. Hassan
Trieste, Italy

Acknowledgements

This book is based on a large scale research project on South-South collaboration in health biotechnology. I am grateful for the opportunity to have worked with my collaborators on this project: Sachin Chaturvedi from India, Wen Ke from China, Victor Konde from Zambia, Magdy Madkour from Egypt and last but not least Tirso Sáenz and Maria Carlota de Souza Paula from Brazil. Our shared interest, complementary expertise and frank and passionate discussion helped make this project successful. The contributions of their colleagues Sidar Abdusamad, Sahar Aly, Nefertiti El-Nikhely, Marwa G. Elwakil, Chen Guang, Zhang Jiuchun, Lexuan Li and, Heba Maram, are also greatly appreciated. I am also deeply thankful to Amitav Rath who read the whole manuscript and provided us with frank and constructive criticism.

This project was funded by grants from Genome Canada through the Ontario Genomics Institute (OGI) and the International Development Research Centre (IDRC) and supported by the McLaughlin-Rotman Centre for Global Health at the University Health Network and the University of Toronto. I want to express gratitude for the financial support and input from these organisations. Special thanks go to David O'Brien and Jean Woo for their advice and comments on the project and Bill Carman for his publishing support. I also want to acknowledge the contribution from the Canadian Institutes of Health Research (CIHR) for supporting me with a New Investigator Award.

My colleagues at the McLaughlin-Rotman Centre for Global Health have provided us with a conducive research environment to conduct this research which I am thankful for. I want to thank Abdallah Daar and Peter

Singer for their contribution to this project. Special thanks go to Christina Melon and Sharon Chakkalackal for their important research input and patiently helping to run this project smoothly. Without Christina's quirky comments this whole endeavour would not have been as fun-filled. Thanks also to my graduate students Andrew Kapoor and Monali Ray who worked hard at various phases of this research. I also greatly appreciate comments provided by other colleagues at the McLaughlin-Rotman Centre on parts of this book. They are in alphabetical order: Jennifer Bell, Jocalyn Clark and Dominique McMahon. Further thanks go to Jennifer Chadder, Jan Cooper, Katherine MacDonald, Michelle Li, Gajen Perinpanayagam and May Sanaee for their assistance.

The firm Science-Metrix conducted scientometric analysis which is widely used throughout the book. Special thanks go to David Campbell, Grégoire Côté, and Éric Archambault. Thanks also to Joseph Wong for his contribution to the case study on China's and India's collaboration with Sub-Saharan Africa. Further thanks go to Cammy Richelli who copy-edited the entire manuscript and to Sigrid Plamenco, Sarah Scott and Sina Zere for their assistance with financial arrangements.

I greatly appreciate that Mohamed Hassan introduced this book by writing a Foreword that puts it in wider context. When planning this project we obtained advice from several experts in developing countries. I particularly want to thank Einard Blanco-Garcia, Claudia Chamas, George Essgeby, Normando Iznaga-Escobar, John Mugabe, Jiunya Liu and Prasit Palittapongarnpim for taking part in a planning meeting at the onset of the project. We also want to thank all the experts we interviewed for this project and who generously shared their expertise and time. Their experience in South-South collaboration was invaluable for this project.

Lastly I want to express my deep gratitude to Christopher Evans for his patient support and reading over endless text excerpts and discussing with me the intricacies of South-South collaboration and to the team at House&Hound for bringing balance to my life.

— Halla Thorsteinsdóttir

Abbreviations

ABLE	Association of Biotechnology Led Enterprises (India)
AMANET	African Malaria Network Trust
AMVTN	African Malaria Vaccine Testing Network
ANDI	African Network for Drugs and Diagnostics Innovation
ANVISA	Agência Nacional de Vigilância Sanitária (Brazil)
API	Active Pharmaceutical Ingredient
ARIF	Average of Relative Impact Factors
ASEAN	Association of Southeast Asian Nations
BAPA	Buenos Aires Plan of Action
BPL	Biotechnology Pharmaceutical (China)
BIPA	Bilateral Investment Promotion and Protection Agreement
BRIC	Biotechnology Regional Innovation Centres (South Africa)
CAS	The Chinese Academy of Sciences (China)
CASTEP	China-Africa Science and Technology Partnership Program
CBAB/CABBIO	Argentinean-Brazilian Centre for Biotechnology
CECMED	Center for Quality Control of Medicines (Cuba)

CIGB	Centro de Ingeniería Genética y Biotecnología (Cuba)
CIM	Centre for Molecular Immunology (Cuba).
CNPq	Conselho Nacional de Desenvolvimento Científico e Tecnológico (Brazil)
COMCGEB	Virtual Centres for Genetic Engineering
COMESA	Market for Eastern and Southern Africa
COMSATS	Commission on Science and Technology for Sustainable Development in the South
CPPI	Centre for Research in Immunological Products (Brazil)
DAAD	Deutscher Akademischer Austauschdienst (Germany)
DBT	Department of Biotechnology (India)
DST	Department of Science and Technology (India, South Africa)
EAC	East African Community
ECOWAS	Economic Community of West African States
EGFR	Epidermal Growth Factor Receptor
EPO	Erythropoietin
ENRA	Egyptian regulatory authority (Egypt)
FDI	Foreign Direct Investment
G-77	UN Group of 77
GAFTA	Greater Arab Free Trade Area
GDP	Gross Domestic Product
GEBRI	Genetic Engineering and Biotechnology Research Institute (Egypt)
GERD	Gross Domestic Expenditure on Research and Development

GNI	Gross National Income
GNP	Gross National Product
HPLC	High Performance Liquid Chromatography
IAIB	India-ASEAN Institute of Biotechnology
IAVI	International AIDS Vaccine Initiative
IBSA	India, Brazil, South Africa initiative
IBSC	Indo-Brazil Science Council
ICDDR,B	International Centre for Diarrhoeal Disease Research (Bangladesh)
ICGEB	International Centre for Genetic Engineering and Biotechnology
IF	Impact Factor
INPI	Brazilian Patent Office (Brazil)
IP	Intellectual Property
ITEC	Indian Technical Cooperation Programme (India)
JE	Japanese encephalitis
KEMRI	Kenya Medical Research Institute (Kenya)
MFN	Most Favoured Nation
MoU	Memorandum of Understanding
mtDNA	Mitochondrial DNA
MuCSAT	Mubarak City for Science and Technology (Egypt)
NEPAD	New Partnership for Africa's Development
NCDEG	National Centre for Diabetes, Endocrinology and Genetics (Jordan)
NIPRID	Nigerian Institute for Pharmaceutical Research and Development (Nigeria)
NMCC	National Malaria Control Centre (Zambia)

NMR	Nuclear Magnetic Resonance
OFDI	Outward Foreign Direct Investment
PID	Integrated Programme on Genetics (Brazil)
PIDE	Integrated Programme on Tropical Diseases (Brazil)
PRONAB	National Biotechnology Programme (Brazil)
R&D	Research and Development
REC	Regional Economic Community
RIF	Relative Impact Factor
SAARC	South Asian Association for Regional Cooperation
SADC	Southern African Development Community
SANBio	Southern Africa Network for Biosciences
SHIP	Shenzhen High-Tech Industrial Park (China)
SSTCC	State Science and Technology Commission of China (China)
TIA	Technology Innovation Agency (South Africa)
TCM	Traditional Chinese Medicine
TDRC	The Tropical Diseases Research Centre (Zambia)
TRIPs	Trade Related Aspects of Intellectual Property Rights
TWAS	The Academy of Sciences for the Developing World
UNDP	United Nations Development Programme
VACSERA	Holding Company for Biological Products and Vaccines (Egypt)
VL	Visceral Leishmaniasis
WHO	World Health Organization
WHSP	Western Havana Scientific Pole (Cuba)
WTO	World Trade Organization

1 Introduction

AUTHORS: *Halla Thorsteinsdóttir, Andrew Kapoor, Sahar Aly, Sachin Chaturvedi, Nefertiti El-Nikhely, Marwa G. Elwakil, Wen Ke, Zhang Jiuchun, Victor Konde, Lexuan Li, Magdy A. Madkour, Heba Maram, Tirso W. Sáenz, Maria Carlota de Souza Paula*

Cooperation between developing countries, or South-South collaboration, has been a frequent topic on the agenda of developing countries since the mid-1950s. As developing countries' profiles change, and some have become major players in the global economy, their international collaboration is likely to change as well. Recently, there has been growing emphasis on collaboration involving science and technology amongst developing countries, and countries in the South are increasingly signing agreements between themselves aimed at fostering their collaboration. Despite the increased focus on such South-South collaboration, not much is known about the extent of South-South collaboration in science-intensive fields, who are the main collaborators or what it contributes to developing countries. In this book, we examine South-South collaboration in one science-intensive field, the health biotechnology sector, in order to understand the opportunities and constraints afforded by South-South collaboration and the factors and conditions that shape the collaborations. It is based on primary research by a team of researchers from Brazil, Canada, China, Egypt, India, South Africa and Zambia and our focus is both on research collaboration mainly involving academic institutions or public research organisations, and entrepreneurial collaboration, typically involving private sector firms. This chapter provides a historical context to our research on South-South collaboration, an introduction to the study, and a description of the methodology used. We start with an overview of South-South collaboration in general, followed by collaborations specifically in science and technology and present key rationales for South-South health biotechnology collaboration. We then present the main objectives of our research, define key terms, and describe the methodology applied in our research on South-South collaboration. At the end of the chapter, we describe the structure of the rest of the book.

1.1 History of South-South collaboration

Many developing countries, particularly those in Africa and the Caribbean, were taking their first steps as independent nations in the late 1950s and early 1960s. The political relations and terms of trade with their former colonial powers had typically been unfavourable to developing countries, but still the bulk of their trade, educational, scientific, and cultural relations were with those countries. With independence, there were political and economic incentives for developing countries to find alternatives to relations with Northern countries and ties with other developing countries were an obvious option (Rath and Lealess, 2000).

The Bandung Conference in 1955 in Indonesia marked the beginning of cultivating formal relations between developing countries. It was a forum which brought a number of Asian and African countries together to talk about their common concerns. Their historical communality and importance of peace was underscored by the President of Indonesia Dr. Ahmad Sukarno, when he said at the onset of the conference:

> All of us, I am certain, are united by more important things than those which superficially divide us. We are united, for instance, by a common detestation of colonialism in whatever form it appears. We are united by a common detestation of racialism. And we are united by a common determination to preserve and stabilize peace in the world... (Government of India, 2005).

The conference announced the importance of closer ties between developing countries and technical cooperation between them in the form of exchanging experts and trainees, sharing equipment, and conducting pilot projects. The Bandung Conference can be seen as the first step in the formation of various South-South groupings such as the Non-Aligned Movement (1961) and the Group of 77 (1964). Several United Nations' organisations also followed and promoted South-South collaboration and the UN organised the United Nations Conference on Technical Cooperation among Developing Countries in Buenos Aires, Argentina in 1978. At this conference a strategic framework for technical cooperation between developing countries was developed, the so-called Buenos Aires Plan of Action (BAPA).

South-South collaboration can be in different formats. It can involve bilateral collaboration between two developing countries. Both China and India, for instance, have had bilateral collaborations involving technical cooperation with a number of African countries since the 1960s, when many of them became independent (UNCTAD, 2010). Regional collaboration initiatives are also common amongst developing countries and typically target both closer political and economic relations. An example of such collaboration is the Association of Southeast Asian Nations, or ASEAN, which was established in 1967, originally between Indonesia, Malaysia, the Philippines, Singapore and Thailand. More recently or in 1985, the South Asian Association for Regional Cooperation (SAARC) was established originally between Bangladesh, Bhutan, India, the Maldives, Nepal, Pakistan and Sri Lanka, with Afghanistan joining later.

Latin American countries have also been active in regional collaboration and Argentina, Brazil, Chile, Mexico, Paraguay, Peru and Uruguay, for example, established the Latin American Free Trade Association (LAFTA) in 1960 with the intention of creating a common market in Latin America. More recently, a subset of these countries, Argentina, Brazil, Paraguay and Uruguay established Mercosur in 1991, to facilitate the movement of goods, people, and currency between them. In Africa there is the African Economic Community (AEC) with several overlapping regional economic blocs. The Common Market for Eastern and Southern Africa (COMESA) which was established in 1993 is an example of such a bloc and has 19 member states, from different parts of Africa. Even though generally the regional associations discussed above have initially focussed on political and economic cooperation, many of them have later incorporated cooperation programmes in science and technology. There are also some multilateral initiatives between developing countries that cross different continents. A recent example of such multilateral initiatives between developing countries is the India, Brazil, South Africa, or IBSA initiative.

It is a developmental initiative for promoting South-South cooperation and exchange in several areas including energy, education, health, trade, and science and technology.

Rath (2001) analyses the history of South-South collaboration and demonstrates how it has gone through periods of growth and decline at different times during the last few decades. He argues that while developing countries recognised the need for such cooperation in the 1960s, they were not successful during those years in establishing significant economic cooperation. In general, developing countries were producing similar products so there was not much scope for their trade. Furthermore, poor transport and communication networks hampered South-South collaboration. According to Rath (2001):

> These early efforts of South-South co-operation did provide the basis for further articulation of the concept and for strengthening such relationships in the 1970s. This decade was marked by great optimism about the ability of the South to reshape the international structure of power and economic relations in a more equitable direction.

As mentioned above, the 1970s saw strategic planning of South-South collaboration with the BAPA action plan. During the 1970s, there was also a build-up of institutional structures to promote South-South collaboration and, for instance, a Special Unit was established in 1978 within the United Nations Development Programme (UNDP) to promote, coordinate, and directly support South-South collaboration both within United Nations' organisations and more widely around the world.

After the enthusiasm for South-South collaboration in 1970s, the 1980s proved to be more challenging for collaborations amongst developing countries. These were difficult times economically for many developing countries as they were going through economic recession and the so-called debt crisis. Developing countries were pressured to repay loans to international creditors under dire terms that they had taken during economic booms in the preceding decades. A further challenge for South-South collaboration in the 1980s was that funding to the UN was cut and thus it could not support the levels of South-South activities planned. After the meagre times of the 1980s, South-South collaboration appears to be building steam in the 1990s and 2000s. Some developing countries have been experiencing significant economic growth, particularly the emerging economies, including Brazil, China and India (OECD, 2009). They play a larger role in the global economy and outsourcing in emerging economies has become a common business practice. Instead of being exporters of primary commodities as they traditionally were, these developing countries now also provide value added high technology services and products for the global economy. Advances in communication and information technology, liberalisation of international trade, and financial regulation have facilitated globalisation and formation of international networks involving developing countries (Sagasti, 2006; Dervis, 2005).

Trade between developing countries has grown extensively. From 1995 to 2005, South-South exports increased by 197 per cent, while its exports to the rest of the world increased by 143 per cent (UNCTAD, 2009). In 2005, total South-South exports were valued at US $1.7 trillion and accounted for 46 per cent of total exports from the South (UNCTAD, 2009). This total is estimated to have grown at an average annual rate of 13 per cent from 1990 to 2008 (Rosales and Kuwayama, 2010). The South-South trade is, however, not evenly spread amongst developing countries. The top traders are the emerging economies, Brazil, China and India, and 70 per cent of the trade is within Asia (UNCTAD, 2009). There has also been a significant increase in South-South foreign direct investment (FDI), and it was estimated to have increased from around US $15 billion in 1995 to around US $46 billion in 2003 and constitute around 35 per cent of total FDI flows in developing countries (Battat and Aykut, 2005). FDI is particularly active within regions with the greatest flow of funds in Asia and Latin America. Southern multinationals have started to appear on the scene since the mid-1980s and are particularly active in infrastructure and extractive sectors. But several pharmaceutical firms from developing countries have become multinational firms, including Cipla (Mumbai) and Dr. Reddy's Laboratories (Hyderabad) from India and Aspen Pharmacare (Durban) from South Africa.

1.2 Drivers of South-South collaboration

South-South collaboration has thus expanded significantly in the last decades and it is driven more by economic factors reflected in increased South-South trade and FDI than before. In the early days of South-South collaboration, ideological drivers were prominent and, as mentioned above, developing countries wanted to create alternatives to ties with their former colonial powers. Developing countries felt they needed to diversify their economies, and decrease their reliance on Northern markets (Rath and Lealess, 2000). A further impetus to diversify their trade is that trade relations with developed countries were commonly perceived to be unfavourable for developing countries. There was, therefore, a motivation to substitute South-North economic linkages with South-South linkages.

All along there has also been a political force fuelling South-South collaboration: a desire by developing countries to stick together to counteract the political influences of the more powerful Northern countries. Jointly, they attempt to influence the international agendas, for example, those of the World Trade Organization and the G20 countries (Harris, 2005). For instance, developing countries pushed successfully for an exception to patent protection mandated by the Trade-Related Aspects of Intellectual Property Rights (TRIPs) agreement to be able to address public health emergencies (see e.g. Cohen *et al.*, 2005). Some South-South initiatives state explicitly their intention to work together to strengthen their global voice. For instance, the IBSA initiative declares that:

> the IBSA Dialogue Forum plays an increasingly important role in the foreign policies of India, Brazil and South Africa. It has become instrumental for promoting ever closer coordination on global issues between three large multicultural and multiracial democracies of Asia, South America and Africa.

and later on expands that the IBSA initiative:

> strengthens the voice of developing countries and their capacity to contribute to global decisions that impact on their populations. The IBSA Forum contributes, therefore, to the goal of a fair and equitable world order. (IBSA, 2006).

Over time, however, the perceived need of developing countries to distance themselves from their former colonial powers seems to have diminished and reference is more frequently being made to the sharing of common needs or clusters of common interests (Gunatilleke, 1993; Rath and Lealess, 2000). Developing countries have to address many environmental, health, and climate-induced problems that developed countries do not suffer from. South-South collaboration affords the possibility of conducting joint research and learning from one another in order to address these problems. There is also a strong need in developing countries to develop low cost solutions to their problems, as large proportions of their populations are poor. These common needs, therefore, provide an added rationale for South-South collaboration.

There is, however, an increasing divergence between developing countries, with some countries experiencing high levels of economic growth, gaining riches, and building impressive capacity in science and technology, while others have limited development. The increasing divide has led to concerns about the increasing South-South divide, 'there is a disturbing emergence of a South-South gap in capabilities between scientifically proficient countries (Brazil, China, India and Mexico, for example) and scientifically lagging countries, many of which are located in sub-Saharan Africa and the Islamic world' (Hassan, 2005). In the health biotechnology sector, the difference between developing countries is certainly growing with some emerging economies becoming lead contributors to the field globally. China, for instance, ranks number two globally in terms of number of health biotechnology publication in international peer-reviewed journals, just behind the United States. Still, the divergence does offer an increasing potential for the scientifically proficient developing countries to share their expertise and promote capacity building in countries lagging behind scientifically and technologically. When developing countries have limited scientific and technological proficiency, there is less scope for them to collaborate. Some difference is, therefore, likely to be an asset and provide impetus for collaboration, but the difference has to be balanced with enough common interest for the collaboration to be of mutual value.

1.3 Increased emphasis on South-South scientific and technological collaboration

In recent years there appears to be a growing emphasis placed on scientific and technological collaboration amongst developing countries. They are increasingly signing scientific and technological agreements with each other and placing more emphasis on meeting social and economic needs with those agreements (Dickson, 2003). China, for example, has signed bilateral agreements incorporating scientific cooperation with 11 African countries, and has set up a Development Fund for Africa which supports cooperative projects between African nations and China (Hassan, 2007a, 2007b). India has also expressed its political commitments to promote collaboration in science, technology, and research and development (R&D) with Africa. It has, for example, launched the Raman fellowship programme and several other fellowships for African nationals with a budget of about US $40 million (India-Africa Forum, 2008a, 2008b). Capacity-building efforts are strong foci in these initiatives and, in many cases, biotechnology and/or health are identified as priority areas for the collaboration. Increasingly, developing countries are also including scientific collaborative components in their multilateral trade and development endeavours and, for instance, NEPAD (New Partnership for Africa's Development), Mercosur, ASEAN, SARRC, and IBSA have been promoting scientific and technological partnerships within their member countries. TWAS, the Academy of Sciences for the Developing World, is a further example of a multi-country initiative promoting South-South collaboration. It is an autonomous, international organisation founded in 1983 that provides support in many scientific areas for South-South collaboration in capacity building and joint research (Schaffer, 2005).

Despite this increasing emphasis on South-South collaboration and growing emphasis on scientific and technological collaboration, there is limited research examining South-South collaboration in scientific and technological fields. It is unknown if this renewed South-South emphasis represents solely political rhetoric or if developing countries are pushing South-South collaboration in health biotechnology to become a reality. The few studies that have focussed on examining South-South scientific collaboration have mostly involved scientometric analysis, for example, of India's and China's co-publications (Arunachalam and Viswanathan, 2008) and of within African co-publications (Boshoff, 2009; Boshoff, 2010). There is therefore full reason to examine further to what extent South-South collaboration has been a reality in science-intensive fields and examine what opportunities, challenges, and impacts it has had so far.

1.4 Potential factors and conditions shaping South-South health biotechnology collaboration

Lack of research on South-South collaboration is in stark contrast to research on international collaboration of developed countries or

North-North collaboration and developing countries' collaboration with developed countries, or South-North collaboration. There is extensive literature on the different aspects of such collaboration (Bradley, 2008), such as motivation for the collaboration (Katz and Martin, 1997), roles of different participants in the collaboration (Gaillard, 1994; Jentsch and Pilley, 2003; Maselli *et al.,* 2006), impacts of the collaboration (Blickenstaff and Moravcsik, 1982; Shrum and Campion, 2000; Shrum and Shenhav, 1995), the ethical and political aspects of the collaboration (Costello and Zumla, 2000; Rakowski, 1993), and how new types of self-organised global networks involving developing countries are being formed (Wagner, 2008). Some of the research findings on South-North collaboration are likely to be applicable to South-South collaboration.

1.4.1 Rationale for research collaboration in health biotechnology

As a number of developing countries, such as Brazil, China, Cuba, India and South Africa, have built up capacity in health biotechnology, the sector affords options for increased South-South research collaboration (Morel *et al.,* 2005; Thorsteinsdóttir *et al.,* 2004). South-South collaboration can potentially strengthen the health research in these leading developing countries, and also promote capacity building in the countries that lag behind. Several arguments can be made for why developing countries should work together on research activities in health biotechnology. Some of the main ones include:

1. *Shared health needs:* Developing countries' health needs have more in common with each other than they do with the health needs of developed countries. As a result, it can be of mutual interest for developing countries to cooperate and carry out health biotechnology research that addresses common health problems. Many communicable diseases such as malaria and leishmaniasis are predominantly a burden of developing countries. Developing countries suffer, in addition, from an increasing non-communicable disease burden and, as their resources are limited, they need effective and affordable ways of addressing this burden.
2. *Expensive research requirement:* There is a need to share expensive instrumentation/infrastructure and other resources required

for medical research. Research on collaboration in developed countries, or North-North collaboration, has indicated that scientists commonly collaborate to share instruments and the costs of R&D (de Solla Price, 1986; Katz and Martin, 1997). It is likely that the need to share research resources is even more acute in developing countries than in developed countries as the former group has more limited resources for science and technology. The policy of establishing centres of excellence in developing countries reflects an emphasis on creating shared infrastructure for research (Dufour, 2002; Sagasti, 2004).

3. *Specialisation of knowledge:* Research on developed countries has shown that increasing specialisation of knowledge calls for science, technology, and innovation to become more collaborative (Faulkner *et al.,* 1995; Katz and Martin, 1997). The complexity of problems now at hand requires different types of expertise which may not be available in any single developing country. Neighbouring developing countries can, therefore, use collaboration as a means of gaining specialised knowledge they might be lacking in the different subfields of health research.

4. *Indigeneous resources:* Traditional knowledge, biodiversity, and populations with specific genomics characteristics can encourage health biotechnology collaboration between countries. Some developing countries may possess traditional knowledge or biodiversity that offers promising health solutions (Motari *et al.,* 2004). Countries may also have small and isolated populations that are valuable for research in genomics and therefore attract international collaboration (Thorsteinsdóttir, 2000). Particular developing countries may have these resources, but lag behind in health research capacity. Collaboration with other developing countries may be a good tool to harness these resources.

Therefore, there are several practical, economic, political, geographic, historical, and cultural reasons for South-South health research collaboration which are reflected in the rationale of bilateral and multilateral efforts to promote collaboration amongst developing countries.

1.4.2 Rationale for entrepreneurial collaboration in health biotechnology

Private sector firms are active contributors to the health biotechnology field and small and medium-sized biotechnology firms in developing countries are increasingly recognised as producers of health innovation aimed at local health needs (Al-Bader *et al.,* 2009; Frew *et al.,* 2008a, 2008b; Frew *et al.,* 2007; Rezaie *et al.,* 2008). Additionally, collaboration has become an integral characteristic of the biotechnology sector (Pisano, 2006; Powell *et al.,* 2005). For health biotechnology companies to be innovative in any country setting, it is of strategic importance that they be able to form partnerships with other firms. There are several potential reasons why developing countries would want to work with each other in health biotechnology. Here we review the main reasons discussed in the literature:

1. *To minimise costs and risks:* Health biotechnology research and developmental activities are characterised by high costs and high risks (Pisano, 2006). The sector deals with complex systems, human bodies, and many lead compounds are rejected after costly clinical trials have been carried out. As such, it can be financially prohibitive for small and medium-sized firms from the South to participate in the more innovative aspects of medical sciences. One way around this obstacle is through collaborative relationships with other companies, in order to lessen the potential financial burden on any one firm. This is particularly important to developing countries as their access to financing, such as venture capital funding, is typically limited (Ferrer *et al.,* 2004; Thorsteinsdóttir *et al.,* 2007).

2. *To access strategic knowledge and technical skills:* Both scientific and product development knowledge in health biotechnology is highly specialised, making it nearly impossible for any one firm or institution to harness it all. Collaborations therefore become a means by which firms can obtain access to a wide spectrum of knowledge, technologies, and skills, and can implement new and relevant findings in their field. The knowledge can be needed for various phases of health development (Hagedoorn, 2002; Lee, 2007). For many small firms that are taking their first steps

in product development, access to regulatory knowledge, for instance, is key to their innovation potential. Gaining access to this expertise through collaboration is typically the only practical means for them to do so.

3. *To access markets:* Alliances between firms are necessary in order to expand their markets and can ease the process of gaining entry into new and foreign markets (Hagedoorn, 2002). Firms in small countries are particularly dependent on exporting their products to survive. Collaborative arrangements with firms in other countries are typically needed in order to obtain this access. Beyond just collaborating with one firm, being able to access foreign distribution channels will enable the product to reach a wider group of potential users than would otherwise be possible, leading to a larger area of competitive advantage, as well as creating greater potential for positive global health impacts.

4. *To address the needs of a common or similar population:* While the aforementioned reasons could be motivating factors for collaborations between any two health biotechnology firms, there are also specific benefits of collaboration for firms from developing nations when collaborating with one another on health R&D. Collaborating regionally and/or with other developing countries translates into a higher likelihood that the populations of interest for the two firms will overlap, have similar genetic qualities, face similar health-related issues, and function in a similar resource environment. These collaborations can thus increase the chance that developing nations find appropriate and sustainable solutions to their health issues.

Apart from identifying motivations for international collaborations, research is also focussing on what needs to be in place in order for the collaboration to lead to innovation. Experts in the field are increasingly stressing the importance of perceiving and promoting international collaboration as a part of a broader innovation systems framework than has been done traditionally (Chataway *et al.,* 2005; Oyelaran-Oyeyinka, 2005; Velho, 2002). Traditionally, South-North scientific and technological collaboration has been promoted in a linear fashion, and the complex interactions and knowledge flows that need to take place, between the various actors in participating countries in order for innovation to take place have been ignored. Motivations for collaborations have been examined or roles of different participants in the collaboration without paying attention to the innovation systems they operate in. Collaboration cannot only be between individuals, firms, or institutions; instead, for innovation to take place, it must be aligned with, and involve, interactions among the larger set of institutions that contribute directly to innovation in the participating countries.

Our research attempts to examine South-South collaboration from an innovation systems perspective in order to understand how we can cultivate innovation through the South-South collaboration. That involves focussing on the wider factors and conditions that shape the collaboration. The innovation systems framework is a systemic approach to understand and influence technological changes and development (Lundvall, 1992; Nelson, 1993; Edquist, 1997). It has become clear that a systemic approach is needed to understand the complex factors and conditions that shape innovation. Innovation systems consist of institutions that contribute to the creation, diffusion, and use of new, useful knowledge, which are held together by a web of linkages and synergies. A recent definition defines an innovation system as being:

> an open, evolving and complex system that encompasses relationships within and between organizations, institutions and socio-economic structures which determine the rate and direction of innovation and competence-building emanating from processes of science-based and experience based learning. (Lundvall *et al.*, 2009: 6)

Innovation systems include both formal organisations such as universities, research centres, health centres, firms, and government, including state policies and regulations such as biosafety and intellectual property (IP) laws; and informal institutions such as social and cultural norms. Interaction among these elements contributes to a process of innovation that is non-linear and multi-directional. Learning and problem solving are central concepts of innovation systems. Learning-by-doing, by-using, and by-interacting among producers and users of knowledge is characteristic of the cumulative and continuous nature of innovation systems (Freeman, 1987). If we actually start to apply the lens of innovation systems to

South-South collaboration, we realise that the collaboration is actually likely to involve interactions of the different innovation systems in the participating countries. We therefore pay particular attention to how systemic alignment encourages or challenges the collaboration.

We specifically want to focus on collaboration in the health biotechnology sector. Our previous research on health biotechnology innovation demonstrated that economies such as Brazil, China, Cuba and India have built up capacity in health biotechnology and it identified several cases where South-South collaboration played a significant role in building capacity and promoting development of the health biotechnology sector (Thorsteinsdóttir *et al.,* 2007; 2004). While the scope in our previous research was to only have a cursory look at the role of South-South collaboration in health biotechnology, in the study discussed in this book we have had the opportunity to go into greater depth in studying South-South health biotechnology collaboration using different data collection methodologies.

1.5 Study on South-South health biotechnology collaboration

We have seen from above that several potential drivers can encourage South-South collaboration. Some of them promote collaboration in research activities, typically carried out by academic institutions, hospitals, or research institutions, whereas others promote collaboration in entrepreneurial activities, typically carried out by private sector firms. In order to gain a better understanding of the role of South-South collaboration in the health biotechnology field a team of researchers from Brazil, Canada, China, Egypt, India and Zambia conducted a large scale research project. The overall goal of the study was to examine the role of South-South collaboration for health biotechnology development in developing countries. More specifically the objectives were to:

1. Map the levels, geographic distributions, and main characteristics of South-South research collaboration as well as entrepreneurial collaboration.

2. Identify the opportunities, reasons, challenges, and impacts of South-South collaborations in health biotechnology and examine the factors and conditions that shape the collaborations.

3. Devise recommendations on how developing countries can cultivate further South-South collaboration in health biotechnology and how these can contribute towards scientific, economic, and health improvements.

We considered those countries to be developing countries which the World Bank classifies as low- or middle-income economies and those countries to be developed countries which the Bank classifies as high-income countries. The World Bank's classification is based on countries' gross national income (GNI) per capita (*www.worldbank.org*) and considers previous developing countries such as South Korea and Singapore to be high-income countries and thus excluded from our study. We then defined collaboration between two low- or middle-income countries to be South-South collaboration, even though some low- or middle-income countries, such as China are located in the Northern hemisphere. Likewise we consider collaboration between a low- or middle-income country and a high-income country to be South-North collaboration. We used a general definition of collaboration proposed by Jane Maienschein which stresses that a collaboration should minimally involve individuals working together toward a common product, and that they have come together in pursuit of a common goal (Maienschein, 1993). This project, focussed on collaboration of individuals working in organisations/firms in developing countries aimed at producing knowledge/products and services in health biotechnology. The project therefore examines a wide scope of South-South collaboration involving both research collaborations and collaborations that focus on the development of products/services in health biotechnology.

1.5.1 Methodology

Our study on South-South collaboration relied on multiple sources of data, including scientometric analysis of co-publications, a survey of firms about their collaborations, and in-depth case studies on bi-national collaboration between particular developing countries. To supplement these data, we used document analysis of policies and background literature, as well as any other statistics or material of relevance to the topic.

Mapping research collaboration

To examine the extent, geographic distribution, and key characteristics of research collaboration we analysed co-publications of researchers from different low- and middle-income countries in health biotechnology and used them as a proxy for collaboration. In this scientometric analysis, South-South research collaboration was defined as co-authored papers by researchers with addresses from more than one low- or middle-income country. For mapping research collaboration we collaborated with members of the firm Science-Metrix (Montréal, Canada) that has specialised in scientometric analysis. Universities and research institutes in the biomedical field typically emphasise publishing in international peer-reviewed journals; therefore examining patterns of health biotechnology publications can give valuable insights into the development of biomedical sciences (Faulkner *et al.,* 1995; Powell and Owen-Smith, 1998; Zucker *et al.,* 1998). Scientometric analysis is sometimes criticised for not being an accurate tool to measure knowledge production in developing countries, because the databases used are more inclined to include journals published by Western countries, typically in English, rather than developing countries' journals. While this is likely to be the case, the coverage of developing countries' journals in these databases has increased considerably in the last few years. We recognise however, the risk that through this methodology we may not have been able to identify all South-South collaborations in health biotechnology but we are still likely to have identified the collaborations that reflect quality research and have had the most international presence. It is also a valuable methodology to look at how collaboration levels are changing over time and also to study what subfields of health biotechnology the collaborative research focusses on.

The Elsevier's Scopus database was used for the scientometric analysis and a subset of all papers published in the 1996–2009 period that fall within the field of 'health biotechnology' was selected. Scopus has a broad coverage of natural and social science literature, including more than 15,000 peer-reviewed journals from over 4,000 international publishers. Scopus makes it possible to identify papers that are co-authored by researchers from institutions in different countries, by linking all the authors of papers to their institutional addresses. Only documents that were peer-reviewed prior to being accepted for publication were retained in the dataset. The document types are mainly articles, conference papers, and reviews, collectively referred to in this book as 'papers'.

The method of selecting health biotechnology papers has been developed by Science-Metrix and perfected over the years. The keywords were selected as follows: firstly, papers were randomly selected from journals specialising in biotechnology; keywords and keyword combinations were then chosen from the titles and author keywords of these papers in order to retrieve other papers in the field of biotechnology. Subsequently, a subset of papers from the biotechnology dataset was built to delineate the domains of health biotechnology. In order to identify papers as being in health biotechnology, a classification scheme of fields and subfields from the United States National Science Foundation applied in its Science and Engineering Indicator and based on the journals that the papers are published in was used. To limit the dataset to papers in health biotechnology, only journals classified in the following subfields of science were included: biomedical research, clinical medicine, and the health sciences. In addition, papers that were found in Medline and that are attached to the search 'meshterm' 'human' were kept in the health biotechnology dataset. Finally, keyword searches (in-title, in-author-keyword, and in-abstract) were performed to identify additional papers related to health within the biotechnology dataset.

To produce datasets on malaria and HIV/AIDs research within health biotechnology, Science-Metrix analysts used a combination of approaches including:

- Keywords searches in the titles, abstracts, and author keywords of papers in the health biotechnology dataset; and
- Retrieval of papers in the health biotechnology dataset that are linked to Medline's search terms related to the 'meshterms' malaria and HIV/AIDs research.

For each dataset, addresses from papers were standardised according to country. As the focus of the study is the analysis of scientific collaboration between developing countries, countries were classified as low- or middle-income economies or high-income economies according to the World Bank classification of countries.

Statistics were produced based on the following indicators:

1. *Number of papers:* Number of health biotechnology papers written by authors associated with countries.
2. *International collaboration:* Number of health biotechnology papers that are co-authored by researcher(s) with address(es) from country A and researcher(s) with address(es) from country B.
3. *Average of relative impact factors (ARIF):* This indicator is a proxy for the quality of the journals in which an entity publishes. Each journal has an impact factor (IF), which is calculated annually based on the number of citations it received relative to the number of papers it published. The IF of papers is calculated by ascribing to them the IF of the journals in which they are published. Subsequently, to account for different citation patterns across fields and subfields of science (e.g. there are more citations in biomedical research than mathematics), each paper's IF is divided by the average IF of the papers in its subfield to obtain a relative impact factor (RIF). For health biotechnology, the subfields are those defined by the National Science Foundation classification and the average IFs are based on the health biotechnology dataset. The ARIF of a given entity is the average of the RIF of papers belonging to it. When the ARIF is above 1, it means that an entity scores better than the world average; when it is below 1, this means that on average, an entity publishes in journals that are not cited as often as the world average.
4. *Network of collaboration among selected countries:* We built a square matrix from the number of papers co-authored by all pairs of selected countries in the period 1996–2009. The software programmes UCINET 6 and NetDraw (Analytic Technologies) were used to process data and produce a representation of the strengths of the relationships between the selected countries. Links between the countries represent collaboration between the two countries involved. The width of the links is proportional

to the number of collaborations between the two entities. These linkages were then superimposed on an image of a world map to display their geographical proximity to one another.

Mapping entrepreneurial collaboration

To map entrepreneurial collaboration we conducted a brief survey on collaborations of health biotechnology/pharmaceutical firms in developing nations. The survey was sent to firms in five developing countries that have been identified as having relatively strong health biotechnology sectors, Brazil, China, Cuba, India and South Africa. The survey was also sent to firms in Egypt to improve the coverage in North Africa and the Middle East. The firms in these countries were asked about their linkages with all other low- and middle-income countries. The survey considered South-South entrepreneurial collaboration to be any work jointly undertaken by firms and organisations in two or more low/middle-income countries that contributes to the production of knowledge, products, or services. Our definition is therefore broad, and involves all types of collaborative activity ranging from marketing to R&D. The logistics of sending the survey to all health biotechnology firms in every developing country was beyond the scope of this project. Instead, the survey was sent to a total of 467 companies in the six countries of focus and the overall response rate to the survey was 62 per cent (see further analysis in Chapter 3).

To identify health biotechnology firms in the selected countries, we referred to biotechnology industry association databases, such as Association of Biotechnology Led Enterprises (ABLE) in India, and other official lists of biotechnology firms in the countries we studied. We collected additional information on the types of biotechnology in which each firm engages—primarily from company websites and publicly available literature—and removed those firms that did not appear to be active in the health biotechnology activities. Such exclusions included: firms that were involved in agribiotechnology, biofuels, or veterinary issues; as well as public institutions or foundations only doing fundamental research, consulting firms, cosmetics companies, and investment or insurance firms. Where a public sector institution was also engaged substantially in health biotechnology entrepreneurial activities, as was the case with some Brazilian and Cuban institutions, they were included in the list of health biotechnology firms. When there was doubt as to whether a firm was active in health biotechnology, three researchers would independently examine information about the firm and come to a consensus. The list that remained included biotechnology firms involved in producing biopharmaceuticals, diagnostics, herbal/natural medicines, nutraceuticals and pharmaceuticals, as well as those that were involved in bioinformatics, clinical trials, contract research, laboratory services, and regenerative medicine.

We sent the surveys on international collaborations to the chief executive officers and heads of research and development and/or partnership development of the health biotechnology firms, as well as other organisations heavily involved in entrepreneurial activities. To encourage a good response rate the survey questionnaire was kept brief, asking a total of 10 questions (for the list of questions, see Box 1.1). When they were engaged in South-South collaborations, we gathered additional information on the country and/or firm name(s) as well as the activities involved, the reasons for the collaborations, the output of the collaborations, who initiated the collaboration, etc. We first administered the survey questionnaire via an online survey service called Instant Survey, to make it easier for the firms to respond. We later made extensive follow-up calls to non-respondents. Firms were then sent their responses for verification. In a few cases, responses from recent in-depth interviews with members of our research team were used in lieu of a survey response.

Our analysis examined the extent of the South-South firm collaboration at the aggregate level compared to South-North collaboration. We also looked at how the survey results varied from country to country and who initiated the collaborations. To visually represent the geographical spread of the linkages, we mapped the findings using social network analysis software, UCINET 6 and NetDraw, to generate the linkages between countries, later superimposing the linkages over an image of the world map. We further looked at several characteristics of the collaboration including the distribution of the activities involved in the South-South collaboration, the reasons for the collaboration, and what outputs the collaborations had led to.

BOX 1.1

Survey questions

1. Does your firm collaborate or work together in any way with firms or organisations in other developing countries? Please see [World Bank link] for the list of countries that fall into this category.

   Choices: Yes or No.

2. If 'YES,' please list the collaboration initiative by presenting the name and location of the partner firm/organisation (e.g., ABC Inc., China).

3. What activities do each of the collaborations entail?

   Choices: R&D, Contract research, Laboratory services, Clinical trials, Manufacturing, Providing supplies, Using supplies, Distribution, Marketing, Training, Other (please specify).

4. What are the reasons for the collaborations?

   Choices: Access to markets, Gain knowledge, Access to technologies/equipment, Access to patients, Access to financing, Provide markets, Provide knowledge, Provide technologies/equipment, Provide patient access, Provide financing, Other (please specify).

5. What types of technologies does each of the collaborations involve?

   Choices: Vaccines, Biopharmaceuticals, Diagnostics, Pharmaceuticals, Drug delivery systems, Herbal medicines, Equipment/devices, Drug delivery systems, Bioinformatics, Other (please specify).

6. What type of partner organisation(s) is/are your firm cooperating with?

   Choices: Firm, Research institute, University, Hospital, Government, Other (please specify).

7. What types of formal arrangements have your firm established with your collaborator(s)?

   Choices: Joint venture, Licensing agreement, Strategic alliance, Subsidiary, Memorandum of understanding, Other (please specify).

8. What types of outputs have each of the collaborations produced?

   Choices: Joint product on market, Joint product in pipeline, Joint patent, Joint publication, Other (please specify).

9. Who initiated the collaboration?

   Choices: Your firm, Partners in developing countries, Your government agency, Your partner's government agency, International organisation, Expatriates, Other (please specify).

10. Does your firm collaborate with firms or organisations in industrialised/high income countries? Please see [World Bank link] for the list of countries that fall into this category.

    Choices: Yes or No.

Case studies on bilateral South-South collaborations in health biotechnology

To understand the opportunities, challenges, and impacts of the South-South collaborations and identify strategies to strengthen the collaborations and their contributions to innovation, we carried out case-study research on collaboration involving 13 developing countries where each case is a bilateral collaboration involving two countries. The selection of our main countries of interest—Brazil, China, India and South Africa—was again based on our previous research that identified these as relatively strong developing countries in the field of health biotechnology (Ferrer *et al.,* 2004; Motari *et al.,* 2004; Thorsteinsdóttir *et al.,* 2004). To improve the coverage in North Africa and the Middle East, we also included Egypt as a focal country. The decision was to base the case-study data collection on existing health biotechnology collaborations to learn from researchers and entrepreneurs that have had direct experience with South-South collaboration. We used our South-South mapping data to identify where research and entrepreneurial collaborations were most frequently, and choose which linkages to examine in more detail. Where possible, we selected linkages that had seemingly little to no involvement from northern partners or institutions. We also considered advice from potential users of our research results when selecting our focus countries. In order to solicit their advice we organised a meeting with potential users of our research at the onset of our project.

The countries we examined in this research were:

Brazil's collaboration with Argentina and Cuba.

China's collaboration with India, Thailand, and Cuba.

Egypt's collaboration with China and Jordan.

India's collaboration with Brazil and Bangladesh.

South Africa's collaboration with Kenya and Zambia.

Sub-Saharan Africa's (i.e., Kenya, Nigeria and South Africa) collaboration with China and India.

For this research we relied heavily on interviews with health biotechnology researchers and entrepreneurs and asked them about their collaboration experiences. In each country we also interviewed other members of the local health biotechnology innovations systems, to gain better insights into the systemic factors that influence the collaboration, the policies and programmes in place to promote South-South collaboration and explore how institutions such as the regulatory and intellectual property regimes impact South-South collaboration in health biotechnology. The individuals interviewed included policymakers who promote collaboration, representatives from drug regulatory agencies, and intellectual property rights experts. Research and entrepreneurial interviewees were asked about their views on South-South collaboration in the field of health biotechnology, and about their specific collaborative projects. The interviews included discussions about the reasons for collaboration; their roles and contributions; and the projects' impacts, etc. Furthermore, we asked questions regarding the factors and conditions that either

Table 1.1

Number of interviewees in the different countries interviewed for case-study research on South-South health biotechnology collaborations

	Countries	Number of interviewees
Brazil study	Brazil	15
	Argentina	17
	Cuba	10
China study	China	20
	India	9
	Thailand	6
	Cuba	3
Egypt study	Egypt	34
	China	16
	Jordan	28
India study	India	24
	Brazil	18
	Bangladesh	8
South Africa study	South Africa	27
	Kenya	10
	Zambia	10
Sub-Saharan African study	Nigeria	34
	Kenya	26
	South Africa	33
Total number	13 countries	348

contribute towards successful collaboration, or have been found to hinder collaborations. Representatives from governments and other relevant institutions in the innovation systems were also asked questions about the possible impacts of the collaboration, and the factors that have hindered or encouraged the collaborations' impacts. Their opinions and suggestions were solicited on how to strengthen South-South collaboration and how it could have a greater impact on improving the health of local populations, as well as how it could strengthen local innovation.

As a part of this study, we interviewed 348 experts in face-to-face interviews in the countries of study. Some of them had several interviewees present at the same time. They were either carried out in the local language or English. In addition, we relied on other sources of data such as relevant background information, policies, statistics, scientometric data, and the examination of the firm survey data. The analysis of the case studies combined descriptive quantitative indicators gleaned from the documents with the in-depth qualitative thematic analysis of the documents and interviews.

1.6 Structure of the book

In the chapters that follow, we present the results and analysis of our research. Chapters 2 and 3 contain results of the mapping exercises, in which, on one hand, we map South-South research collaboration and, on the other hand, map South-South entrepreneurial collaborations. We present information on the levels, geography, and key characteristics of the collaboration from the scientometric and survey research. We also compare the key collaborating countries. In Chapters 4 to 8 we discuss the results from the in-depth bilateral case studies. We organise each chapter according to key focal country examined in this study—Brazil, China, Egypt, India and South Africa—and then discuss their bilateral collaboration in the select countries listed above. For each chapter we review the governmental interest in and support for South-South collaboration, map the focal countries key linkages, discuss in detail the main reasons, challenges, and impacts of their research *versus* entrepreneurial collaboration, and present the main conclusions on the case-study research. In Chapter 9 we also discuss case-study research but focus specifically on China and India's collaboration with sub-

Saharan African countries. We provide an overview of Chinese and Indian governmental interest in and support for sub-Saharan African collaboration, map their key linkages, and discuss in detail the main reasons, challenges, and impacts of their research *versus* entrepreneurial collaboration. In addition, we compare China's and India's roles on the sub-Saharan African continent. In the last chapter, Chapter 10, we take stock of our key learning about South-South collaboration and present recommendations on what can be done to strengthen collaboration amongst developing countries in health biotechnology and enhance the collaborations' contribution towards scientific capacity, economic growth, and health.

References

Al-Bader, S., S.E. Frew, I. Essajee, V.Y. Liu, A.S. Daar and P.A. Singer (2009). "Small but tenacious: South Africa's health biotech sector", *Nature Biotechnology* 27(5): 427-445.

Arunachalam, S. and B. Viswanathan (2008). "South-South cooperation: The case of Indo-Chinese collaboration in scientific research", *Current Science* 95(3): 311 313.

Battat, J. and D. Aykut (2005). "Southern multinationals a growing phenomenon", *Foreign investment advisory service*. The World Bank and International Finance Corporation.

Blickenstaff, J. and M.J. Moravcsik (1982). "Scientific output in the third-world", *Scientometrics* 4(2): 135-169.

Boshoff, N. (2009). "Neo-colonialism and research collaboration in Central Africa", *Scientometrics* 81(2): 413-434.

———. (2010). "South South research collaboration of countries in the Southern African Development Community (SADC)", *Scientometrics* 84(2): 481-503.

Bradley, M. (2008). "On the agenda: North South research partnerships and agenda-setting processes", *Development in Practice* 18(6): 673-685.

Chataway, J., J. Smith and D. Wield (2005). "Partnerships for building science and technology capacity in Africa: Canadian and UK experience", *Africa-Canada-UK exploration: Building science and technology capacity with African partners* 30.

Costello, A. and A. Zumla (2000). "Moving to research partnerships in developing countries", *British Medical Journal* 321(7264): 827.

de Solla Price, D.J. (1986). *Little science, big science... and beyond*. New York: Columbia University Press.

Dervis, K. (2005). *A better globalization: Legitimacy, governance, and reform*. Washington, D.C.: Center for Global Development.

Dickson, D. (2003). "South-South collaboration picks up steam", *SciDev.Net*, November 17.

Dufour, P. (2002). "Knowledge, innovation and development: Enhancing Canada's standard of living through emerging global partnerships", Solicited paper submitted to the Conference Board of Canada for TD Forum on *Canada's standard of living*. Ottawa: IDRC.

Edquist, C. (1997). *Systems of innovation: Technologies, institutions and organizations*. London and Washington, DC: Pinter.

Faulkner, W., J. Senker and L. Velho (1995). *Knowledge frontiers: Public sector research and industrial innovation in biotechnology, engineering ceramics and parallel computing*. Clarendon.

Ferrer, M., H. Thorsteinsdóttir, U. Quach, P. A. Singer and A.S. Daar (2004). "The scientific muscle of Brazil's health biotechnology", *Nature Biotechnology* 22: DC8-DC12.

Freeman, C. (1987). *Technology policy and economic performance – Lessons from Japan*. London: Pinter Publishers.

Frew, S.E., H.E. Kettler and P.A. Singer (2008a). "The Indian and Chinese health biotechnology industries: Potential champions of global health", *Health Affairs (Millwood)* 27(4): 1029-1041.

Frew, S.E., R. Rezaie, S.M. Sammut, M. Ray, A.S. Daar and P.A. Singer (2007). "India's health biotech sector at a crossroads", *Nature Biotechnology* 25(4): 403-417.

Frew, S.E., S.M. Sammut, A.F. Shore, J.K. Ramjist, S. Al-Bader and R. Rezaie (2008b). "Chinese health biotech and the billion-patient market", *Nature Biotechnology* 26(1): 37-53.

Gaillard, J.F. (1994). "North-South research partnership: Is collaboration possible between unequal partners?", *Knowledge, Technology & Policy* 7(2): 31-63.

GoI (2005). *Bandung: Celebrating the fifieth anniversay of the Asian-African conference, 1955*. Delhi: External Publicity Division, Ministry of External Affairs.

Gunatilleke, G. (1993). *The third world in an undivided world facing the challenge : Responses to the report of the South Commission*. London: South Centre. pp. 246-256.

Hagedoorn, J. (2002). "Inter-firm R&D partnerships: An overview of major trends and patterns since 1960", *Research Policy* 31(4): 477-492.

Harris, J. (2005). "Emerging third world powers: China, India and Brazil", *Race & Class* 46(3): 7-27.

Hassan, H.A. (2005) "Small things and big changes in the developing world", *Science* 309 (5731): 65-66.

Hassan, M.H. (2007a) "Building capacity in the life sciences in the developing world", *Cell* 131(3): 433-436.

Hassan, M.H.A. (2007b). "A new dawn for science in Africa", *Science* 316(5833): 1813.

IBSA (2006). "First IBSA summit declaration", Paper presented at the India, Brazil, South Africa Summit, Brasilia.

India-Africa Forum Summit (2008a). *Delhi declaration*. New Delhi, April.

————. (2008b). *India-Africa framework for cooperation*. New Delhi, April.

Jentsch, B. and C. Pilley (2003). "Research relationships between the South and the North: Cinderella and the ugly sisters?", *Social Science & Medicine* 57(10): 1957-1967.

Katz, J.S. and B.R. Martin (1997). "What is research collaboration?", *Research Policy* 26(1): 1-18.

Lee, C.W. (2007). "Strategic alliances influence on small and medium firm performance", *Journal of Business Research* 60(7): 731-741.

Lundvall, B.Å. (1992). *National systems of innovation: Towards a theory of innovation and interactive learning*. London: Pinter.

Lundvall, B.Å., J. Vang, K.J. Josepth and C. Chaminade (2009). "Innovation system research and developing countries", in B.Å. Lundvall, K.J Joseph, C. Chaminande and J. Vang (eds.), *Handbook of innovation systems and developing countries: Building domestic capabilities in a global setting.* Cheltenham: Edward Elgar. pp.1-30

Maienschein, J. (1993). "Why collaborate?", *Journal of the history of biology* 26(2): 167-183.

Maselli, D., J.A. Lys and J. Schmid (2006). "Improving impacts of research partnerships", *Geographica bernensia,* Swiss Commission for research partnerships with developing countries, KFPE, Bern.

Morel, C.M., T. Acharya, D. Broun, A. Dangi, C. Elias, N.K. Ganguly (2005). "Health innovation networks to help developing countries address neglected diseases", *Science* 309(5733): 401.

Motari, M., U.Quach, H. Thorsteinsdottir, D.K. Martin, A.S. Daar and P.A. Singer (2004). "South Africa - Blazing a trail for African biotechnology", *Nature Biotechnology* 22: DC37-DC41.

Nelson, R.R. (1993). *National innovation systems: A comparative analysis.* US: Oxford University Press.

OECD (2009). *Globalisations and the emerging economies: Brazil, Russia, India, Indonesia, China and South Africa.* Paris: Organisation for Economic Cooperation and Development. p.454.

Oyelaran-Oyeyinka, B. (2005). "Partnerships for building science and technology capacity in Africa", *Africa-Canada-UK exploration: Building science and technology capacity with African partners,* 30.

Pisano, G.P. (2006). *Science business: The promise, the reality, and the future of biotechnology.* Boston: Harvard Business School Press.

Powell, W.W. and J. Owen-Smith (1998). "Universities and the market for intellectual property in the life sciences", *Journal of Policy Analysis and Management* 17(2): 253-277.

Powell, W.W., D.R. White, K.W. Koput and J. Owen-Smith (2005). "Network dynamics and field evolution: The growth of interorganizational collaboration in the life sciences", *AJS* 110(4): 1132-1205.

Rakowski, C.A. (1993). "The Ugly scholar: Neocolonialism and ethical issues in international research", *The American Sociologist* 24(3): 69-86.

Rath, A. (2001). "South South co-operation in science and technology for development", *Twelfth session of the high level committee on the review of technical co-operation among developing nations.* Policy Research International Inc.

Rath, A. and S. Lealess (2000). "The forum on South-South cooperation in science and technology", *The forum on South-South co-operation in science and technology: An overview document.* Policy Research Inc., UNDP Seoul Conference, Republic of Korea (14-17 February), May. Seoul: Policy Research International Inc.

Rezaie, R., S.E. Frew, S.M. Sammut, M.R. Maliakkal, A.S. Daar and P. A. Singer (2008). "Brazilian health biotech--fostering crosstalk between public and private sectors", *Nature Biotechnology* 26(6): 627-644.

Rosales, O. and M. Kuwayama (2010). "South-South cooperation: Latin America and the Caribbean", *International trade forum: Emerging markets & the future direction of South-South trade*(2): 24-25.

Sagasti, F. (2006). "Rethinking technical cooperation among developing countries (TCDC) and South-South Cooperation (SSC)", *An issues paper.* Lima.

———. (2004). *Knowledge and innovation for development: The Sisyphus challenge of the 21st century*. Edward Elgar Publishing.

Schaffer, D. (2005). "TWAS at 20, A history of the Third World Academy of Sciences", (Journal Article).

Shrum, W. and P. Campion (2000). "Are scientists in developing countries isolated?", *Science Technology & Society* 5(1): 1.

Shrum, W. and Y. Shenhav (1995). "Science and technology in less developed countries", *Handbook of Science and Technology Studies*. pp.627-651.

Thorsteinsdóttir, H., U. Quach, A.S. Daar and P.A. Singer (2004). "Conclusions: Promoting biotechnology innovation in developing countries", *Nature Biotechnology* 22: DC48-DC52.

Thorsteinsdóttir, H., P.A. Singer and A.S. Daar (2007). "Innovation cultures in developing countries", *Comparative Technology Transfer and Society* 5(5): 178-201.

Thorsteinsdóttir, O.H. (2000). "External research collaboration in two small science systems", *Scientometrics* 49(1): 145-160.

UNCTAD (2010). *Economic development in Africa report 2010: South-South cooperation Africa and the new forms of development partnerships*. New York & Geneva: United Nations Conference on Trade and Development.p. 128

———. (2009). *South-South trade: The realty check*. Geneva: United Nations Conference on Trade and Development.

Velho, L. (2002). "North-South collaboration and systems of innovation", *North-South Research Cooperation*, pp.25-49.

Wagner, C.S. (2008). *The new invisible college: Science for development*. Brookings Institution Press.

Zucker, L.G., M.R. Darby and M.B. Brewer (1998). "Intellectual human capital and the birth of the US bioechnology enterprises", *The American Economic Review* 88(1): 290-306.

2 South-South Research Collaboration in Health Biotechnology

AUTHORS: *Halla Thorsteinsdóttir, David Campbell, Grégoire Côté, Éric Archambault, Abdallah S. Daar, Peter A. Singer*

2.1 Introduction

Scientific scholarship has reflected developing countries' dependence on Northern relations, where most roads have led to Western Europe and North America. Now as more developing countries have built capacity in science and technology, there is the potential for developing countries to lessen their dependence on the North and forge linkages with each other. Developing countries are increasingly signing agreements that encourage scientific and technological collaboration amongst themselves (Hassan, 2007a, 2007b; Rath and Lealess, 2000). These are both bilateral and multilateral initiatives aimed at encouraging tighter linkages amongst researchers in developing countries. As we have discussed in Chapter 1, several reasons can drive research collaboration amongst health biotechnology researchers and they may, for example, work together to share infrastructure and other resources, to gain access to each other's expertise and to biological samples, and to strengthen their potential to produce knowledge that can be used to address their shared health problems (Bozeman and Corley, 2004; Hardy *et al.,* 2008; Katz and Martin, 1997).

Despite efforts to increase South-South research collaboration, not much is known about the extent and characteristics of the linkages that already exist. It is not clear whether South-South research collaboration is rhetoric or reality. There is, for example, no empirical research that examines the levels and patterns of South-South collaboration in health biotechnology; whether or not this type of research is on the rise; which countries are most active in the collaboration; and what effects this research has on the health biotechnology sectors of the participating countries. This deficiency makes it difficult to know if current initiatives to promote South-South collaboration are successful, and to make recommendations on how to structure the initiatives so that they are most effective. This chapter aims to address these gaps in knowledge by presenting results of an analysis of developing countries' publications in the health biotechnology field.

We examined research collaboration between developing countries through a scientometric analysis of health biotechnology publications (See Chapter 1 for fuller discussion of the methodology). Universities and research institutes, both in developed and developing countries, typically emphasise publishing in peer-reviewed journals as a way to share the products of their research and scholarship (Faulkner *et al.,* 1995; Powell and Owen-Smith, 1998; Thorsteinsdóttir *et al.,* 2006; Zucker *et al.,* 1998). By examining patterns of health biotechnology publications can, therefore, provide valuable insight into the characteristics of science in developing countries. Scientometric analysis is sometimes criticised for not being an accurate tool to measure knowledge production in developing countries, because the databases used tend to ignore developing countries' journals. While we will not argue against the potential of there being a bias against developing countries, the coverage of developing countries' journals in these databases has increased considerably in the last few years. There is still some risk that through this methodology we may not have been able to identify all South-South collaborations in health biotechnology, but we are likely to have identified the collaborations that had the most international presence. Scientometric analysis is also a valuable methodology to look at how collaboration levels are changing over time and to study what subfields of health biotechnology the collaborative research focusses on.

We used the Scopus database, and selected a subset of all papers published in the 1996–2009 period that fell within the field of 'health biotechnology.' Co-authored papers with addresses from more than one low- or middle-income country were used as a proxy for South-South research collaboration. We refer to such publications as South-South co-authored papers. In comparison, we refer to papers published by researchers with addresses from both a high-income (i.e., a developed country) and a low- and or middle-income country, as South-North co-authored papers. We start this analysis by looking at the extent of the collaborations, and then examine which countries are most active in South-South collaboration. We then examine characteristics of the collaboration and identify the subfields of health biotechnology that are typically targeted for the collaboration, and look at South-South research collaboration on selected diseases that are particularly relevant to developing countries. At the end we focus on what have been the main impacts of the collaboration so far. This chapter provides a general background at the aggregate level to South-South research collaboration to which the case-study research chapters in this book can be compared.

2.2 Extent of collaborations

During the period 1996 to 2009 almost 1.2 million papers were published in health biotechnology. Of those, 17 per cent had authors from developing countries, whereas 80 per cent had authors from developed countries.[1] It is clear that there is a gap between developed and developing countries' contribution to the field and a relatively small percentage of the papers in this field include authors from developing countries. Still Figure 2.1 shows that the authorship of developing country researchers has more than quadrupled during this period, from 6,190 health biotechnology papers in 1996 to 26,648 papers in 2008, indicating a growing contribution to the field from authors in developing countries. In the same period the increase in health biotechnology papers published by authors in developed countries was only from 61,170 to 76,746 papers.

1. Note that a small portion of the papers in the database did not include addresses for the authors.

FIGURE 2.1

Comparing the number of health biotechnology papers over time published by authors from developing versus developed countries

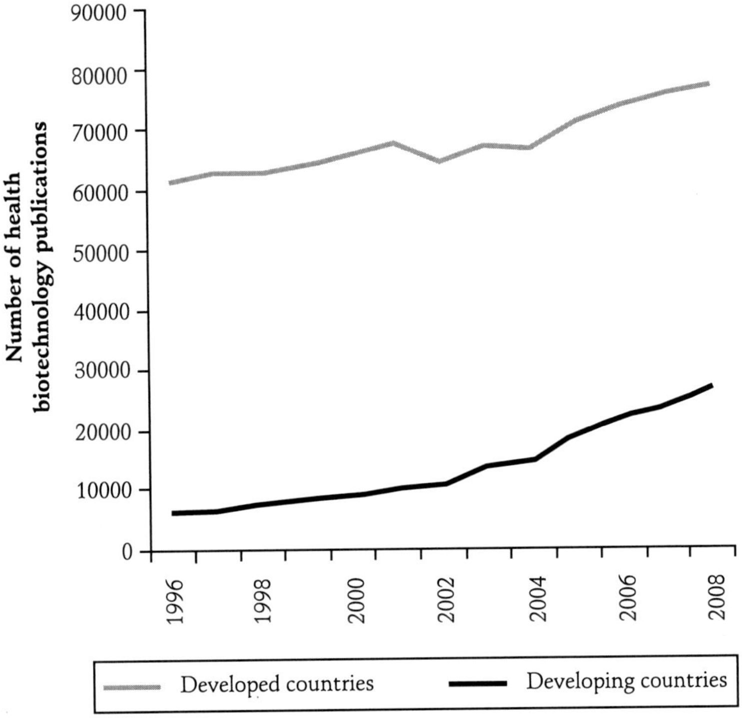

Source: Authors' presentation of data compiled by Science-Metrix using Scopus.

When we looked at which developing countries were increasing their contributions to the health biotechnology field, we saw that in particular, the contributions of Chinese authors grew extensively during this period. They published just over a quarter of developing countries' papers in this field in 1996, but by 2008, they were publishing half of all developing countries' papers. Indian and Brazilian authors were also active in this field, publishing over 10 per cent and 8 per cent, respectively, of developing countries' health biotechnology papers during this time. From this, we can see that the contributions of developing countries to the field are small, but rapidly growing, especially by China. China's growth in health biotechnology publications has been phenomenal. It was in eighth place globally in terms of the number of health biotechnology publications for the period 1998–2001 but by 2006–2009 it was in second place just after the United States.

We calculated the proportion of South-South co-authored papers and South-North co-authored papers to examine how actively developing countries collaborate with each other in the health biotechnology sector (Figure 2.2). The data reveal that almost all developing countries' collaboration is with developed countries and less than 10 per cent of the collaboration is with other developing countries. In comparison South-North collaborations are 97 per cent to 98 per cent of their international collaboration. Note that South-North and South-South collaborations add up to more than 100 per cent because some of the publications involve both types of collaborations. Further it is clear from looking at Figure 2.2

FIGURE 2.2

Comparing the number of health biotechnology papers over time published by developing countries' authors in South-North versus South-South collaboration

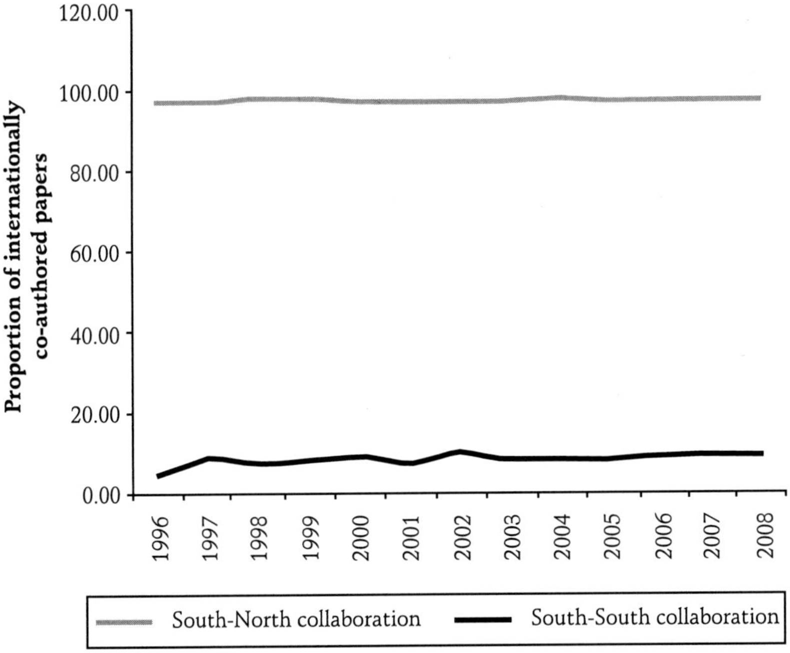

Source: Authors' presentation of data compiled by Science-Metrix using Scopus.

that neither South-North nor South-South collaboration levels seems to have changed during the period studied. Thus, the increased contribution of developing countries to the field of health biotechnology is not due to growing international collaboration, but instead may reflect a strengthened capacity that allows them to publish without international collaboration. Early in their development, developing countries were likely dependent on collaboration with developed countries to complete research, but it is possible that the research infrastructure and capacity in several developing countries have strengthened to a point where this assistance is less frequently needed.

2.3 Geography of South-South collaboration

Aggregate results on the level of South-South collaboration may mask active research linkages between subsets of developing countries. Even though developing countries as a whole do not seem to be active in South-South research collaboration in the health biotechnology field, some countries may be pursuing it more heavily than others. We therefore explored the patterns of collaboration at the country level and present data on the top 10 countries that had the most South-South co-authored papers published between 1996 and 2009 (Figure 2.3). Brazil has the highest number of South-South co-authored papers, with 1,021 publications during the period, and a steady increase in collaborative papers over the years. China has the second most, with 973 South-South co-authored papers; Russia is in third place with 882 papers; and India is in fourth place with 604 papers.

Interestingly, the emerging economies, China, Brazil and India, have the steepest increase in South-South co-authored papers over the latter part of the period (Figure 2.4). As they are also the countries with the strongest economic growth in the last 10 years, this may suggest a connection between South-South collaboration and economic growth, potentially boding well for future South-South collaborative activities in health biotechnology. The steep increase in China's South-South collaboration during the 2006–2009 period is particularly noteworthy and it has surpassed Brazil in the number of South-South co-authored papers during this period. The contribution of Russia, on the other hand, has plateaued and it does not seem to be emphasising much South-South collaboration in the health biotechnology field.

FIGURE 2.3

Number of papers in South-South health biotechnology collaboration published by authors in the countries with the most active collaborations

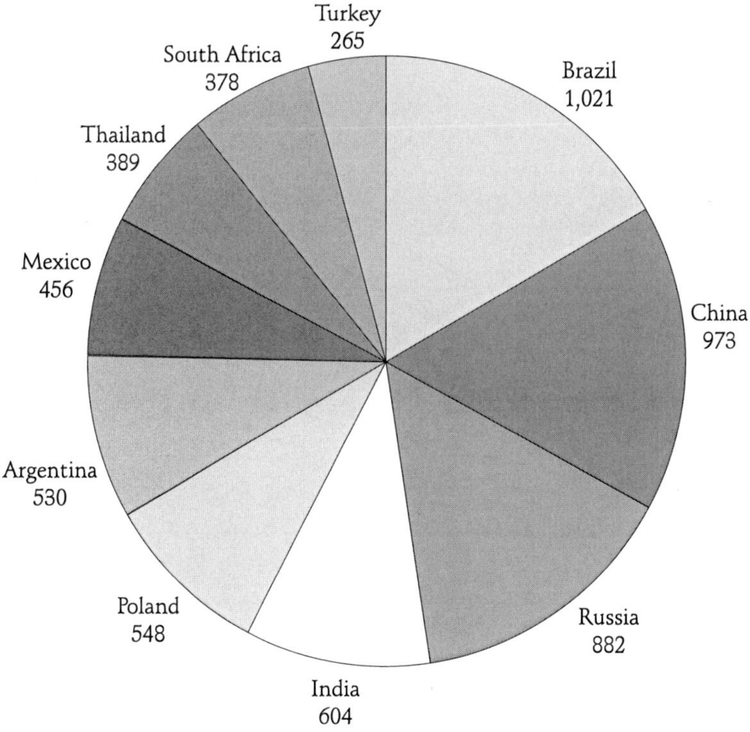

Source: Authors' presentation of data compiled by Science-Metrix using Scopus.

FIGURE 2.4

Number of papers in South-South health biotechnology collaboration published by authors in the countries with the most active collaborations

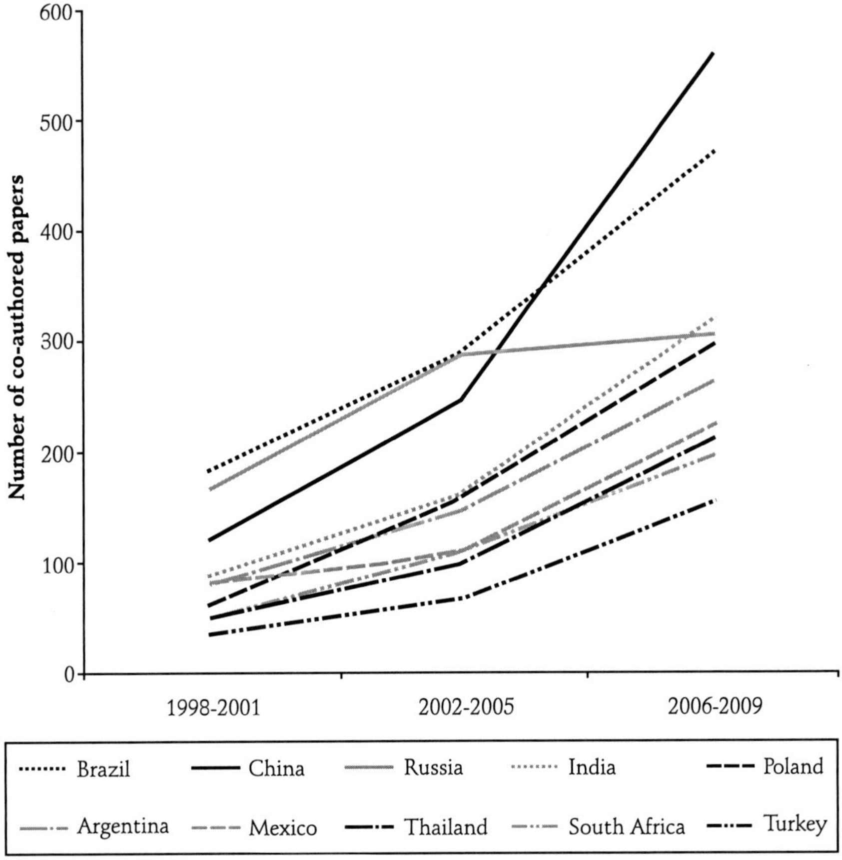

Source: Authors' presentation of data compiled by Science-Metrix using Scopus.

To further explore the different countries' emphasis on South-South collaboration, we looked at the proportion of South-South health biotechnology collaboration of the focal countries of this book *versus* all their international collaborations (Figure 2.5). What is noteworthy from the examination is that all the countries, except China, place a much larger emphasis on South-South collaboration than developing countries in general do. The percentage of South-South co-authored papers *versus* all internationally co-authored papers ranges from only 5 per cent for China to 25 per cent to South Africa, with a mean of 17.4 per cent in the last period for the four countries (excluding China) that we included in our further case-study research. South Africa, Brazil and Egypt are the countries that have the highest percentage of South-South co-authored papers in the last period studied. China places relatively low emphasis on South-South collaboration in the health biotechnology field compared to all of its international collaborations. As shown above, China is by far the largest developing country contributor to the field, and because its South-South collaboration is low, it suppresses the observed rate of South-South collaboration generally for developing countries. A further observation from Figure 2.5 is that South-South collaboration in health

FIGURE 2.5

Proportion of South-South collaboration of all international collaboration for focal countries, 1998–2009

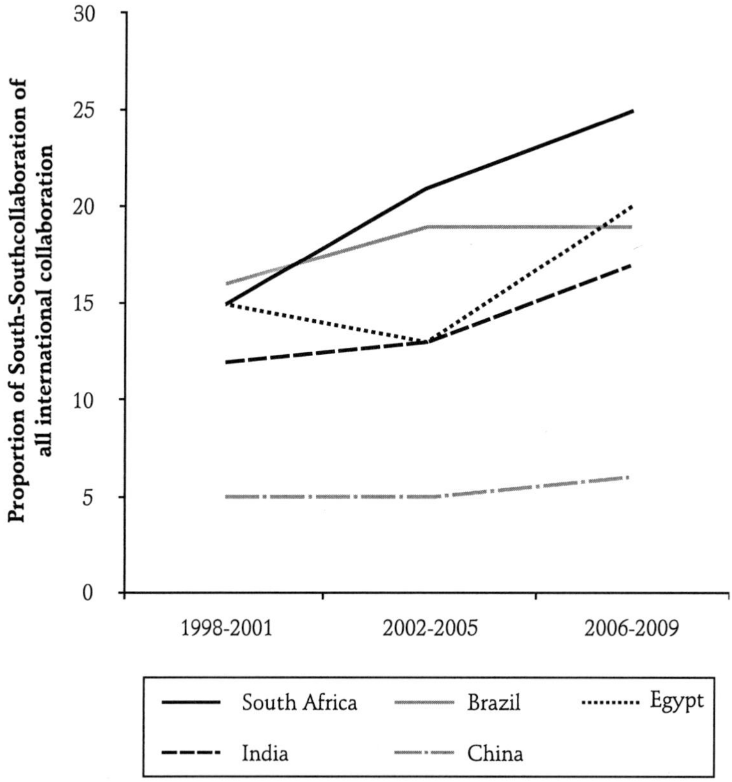

Source: Authors' presentation of data compiled by Science-Metrix using Scopus.

biotechnology appears to be increasing. All of the countries have increased their emphasis on South-South collaboration at the end of the period we studied. Still China showed a small increase from 5 per cent of papers being in South-South collaboration at the beginning of the period to 6 per cent at the end.

To further explore South-South collaborations, we mapped the linkages between developing countries (Figure 2.6). It is clear that there are active South-South collaborations involving Latin American and Asian countries but relatively few linkages involving Africa. Brazil collaborates actively with Argentina and other Latin American countries such as Venezuela, Mexico, Colombia and Chile. Collaboration between Brazil and Argentina has been heavily promoted by both governments to jointly stimulate their biotechnology development. As discussed in the chapter on Brazil's South-South collaborations, the two governments created the Argentinean-Brazilian Centre for Biotechnology (CBAB/CABBIO) in 1987, and have supported research, training, and exchange opportunities for their researchers with each nation allocating annual funding to the programme.

Brazil is the largest producer of health biotechnology publications in Latin America with over 16,000 papers from 1996 to 2009; Mexico ranks second with almost 5,900 papers; and Argentina is third, with around 4,800 papers in the field. Even though Mexico is a relatively large producer of health biotechnology publications, its collaboration with Brazil is more modest than that of Argentina. It seems that the political emphasis Brazil and Argentina have placed on working together has led to an increase in the publication of co-authored papers. Brazil is also linked to other strong countries in health biotechnology, such as China, India and South Africa. This may reflect a strategy to connect with other developing country powerhouses in this field and in the case of India and South Africa it may also reflect collaboration promoted by the IBSA network (see Chapters 4, 6, 8, and 9 for further discussion of the IBSA network).

China has a significant network and its strongest ties are with India—another relatively strong country in health biotechnology—and with Thailand (see case study in Chapter 5 on China's South-South collaborations). What is surprising from looking at Figure 2.6, is how relatively limited India's collaboration is with other developing countries,

FIGURE 2.6

South-South research collaboration network in health biotechnology, 1996–2009

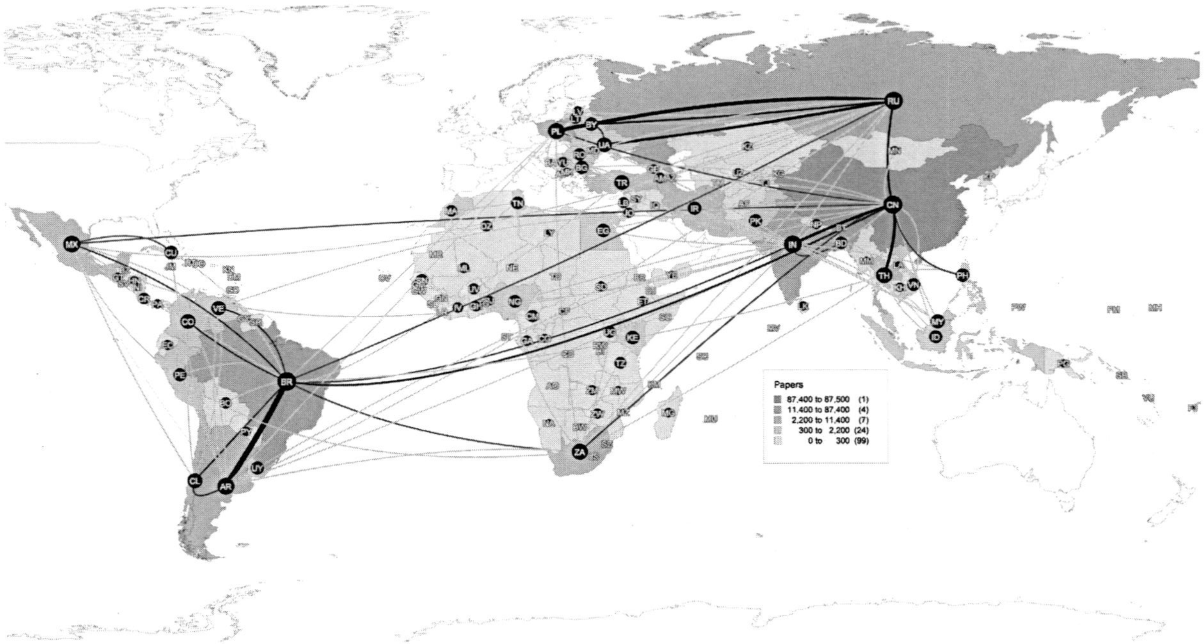

Note: The width of the lines represents the number of collaborations between two linked countries. To ease the representation of partnerships, only linkages of 14 or more co-publications were included on the map.

Source: Compiled by Science-Metrix using Scopus.

considering that it is strong in health biotechnology. In the South-South entrepreneurial collaboration chapter (Chapter 3), we see that India has relatively active firm collaborations, in contrast to the lower levels of research collaborations seen here. Indian firms, for example, frequently collaborate with South Africa, but only modest research collaboration seems to be occurring between the two countries.

What is also noteworthy about India from Figure 2.6, is its collaboration with Bangladesh, a low-income country with a relatively weak standing in the health biotechnology field (see Chapter 7). At the beginning of the period studied, India had almost no collaboration with Bangladesh, but by the end, Bangladesh had become India's main collaborator. From examining the co-authored papers between India and Bangladesh, we can see that almost all the papers are between Indian organisations and the International Centre for Diarrhoeal Disease Research (Dhaka, Bangladesh), which is an international initiative of about 55 donor countries, the Government of Bangladesh, UN specialised agencies, etc., focussed on seeking solutions to diarrheal disease. Also, the emphasis in the papers is on cholera research, a significant health problem in both countries. Therefore, while the India-Bangladesh collaboration reflects a focus on a shared health problem, it involves an international organisation to help address it. The fact that an international organisation is a major partner in the research collaboration between India and Bangladesh may reflect the need to get outside support to fund the collaboration. The need for South-South research collaborations may be generally greater than what we see in our analysis on co-authored papers, as funding issues may be posing an obstacle in the absence of involvement by international organisations and the richer developed countries.

Another relatively strong developing country in health biotechnology, Cuba, is also loosely connected to the South-South research network in health biotechnology. As the United States has had a trade embargo against Cuba since the early 1960s, and generally dominates the field of health biotechnology with a large number of journals in the field based in the United States, it is likely that Cuba's linkages in health biotechnology cannot be accurately measured by examining co-publications in the international scientific peer-reviewed journals. As we note in Chapter 3 on South-South entrepreneurial collaboration, Cuba is more actively involved in South-South entrepreneurial collaboration than in South-South research collaboration.

2.4 Characteristics of South-South research collaboration

The results so far discuss the levels and distribution of the research collaborations in health biotechnology between developing countries. To gain a deeper understanding of this, we explored some of the characteristics of the collaborations and looked at the subfields of health biotechnology of the co-authored papers, examined the linkages of research focussed on HIV/AIDs and malaria, and evaluated the impacts of the collaborations on the visibility of the health biotechnology research.

2.4.1 A strong focus on genetics and heredity

We classified all the South-South co-authored papers into subfields of health biotechnology to analyse them further (see Chapter 1 for the details of the methodology) and present the 10 subfields with most South-South collaboration (Table 2.1). 'Genetics and Heredity' is by far the most common subfield of South-South health biotechnology collaboration. This is the same pattern as observed in North-North collaboration. Even though the subfield of 'Biochemistry and Molecular Biology' trumps 'Genetics and Heredity' in the number of publications from authors in developing countries; only 2 per cent of their papers in the former subfield involve international collaboration *versus* over 5 per cent of the latter. As a result, developing countries have almost 1,200 papers in South-South collaboration in 'Genetics and Heredity' *versus* only around 530 papers in such collaboration in 'Biochemistry and Molecular Biology.' The relatively high rate of collaboration in 'Genetics and Heredity' is likely to reflect the emphasis on international collaboration that has been promoted within the Human Genome Project and other international initiatives taking place in genomics. These projects are fuelled by the belief that in order to understand human diseases better, the environmental factors that trigger disease, and what interventions (e.g. medications) might be effective, there is a need for large scale initiatives that cross several national borders. A high rate of collaboration in 'Genetics and Heredity' also reflects the need to obtain samples from different populations around the globe, and to compare them when identifying the genetic causes of diseases.

TABLE 2.1

Number of South-South collaboration papers in subfields of health biotechnology, 1996–2009

Subfield of health biotechnology	Number of papers
Genetics & heredity	1191
Biochemistry & molecular biology	533
Biomedical engineering	419
Microbiology	399
Immunology	335
Virology	263
Tropical medicine	253
General biomedical research	232
Parasitology	220
Cancer	149

Source: Authors' presentation of data compiled by Science-Metrix using Scopus.

To explore these linkages in the subfields of health biotechnology further, we identified the top 10 countries that collaborate in select subfields (Figure 2.7). We chose 'Genetics and Heredity,' as it is the most common subfield of South-South collaboration; 'Tropical Medicine and Parasitology' (combined), as these two subfields are likely to have an exclusive relevance to developing countries' health situations; and 'Virology,' as it is highly relevant to the burden of HIV/AIDS affecting many developing countries.

Collaboration in 'Genetics and Hereditary' is dominated by the largest countries, with Brazil, China, India and South Africa all amongst the top collaborators. The high rate of South-South co-authored papers in genomics likely reflects the fact that international forces have stimulated collaboration in this field. China was, for example, the only developing country that took part in the Human Genome Project, and likely as a result, is the developing country that has collaborated the most in genomics. We also observe cases where neighbouring countries collaborate in 'Genetics and Heredity' with each other. For example, China collaborates with Mongolia in examining polymorphism in ethnic populations in Northern China (Kong *et al.,* 2003), and South Africa collaborates with Zimbabwe and Tanzania in examining a gene for drug metabolising enzymes in their populations (Dandara *et al.,* 2004). This

FIGURE 2.7

The top countries engaged in South-South research collaboration in selected subfields

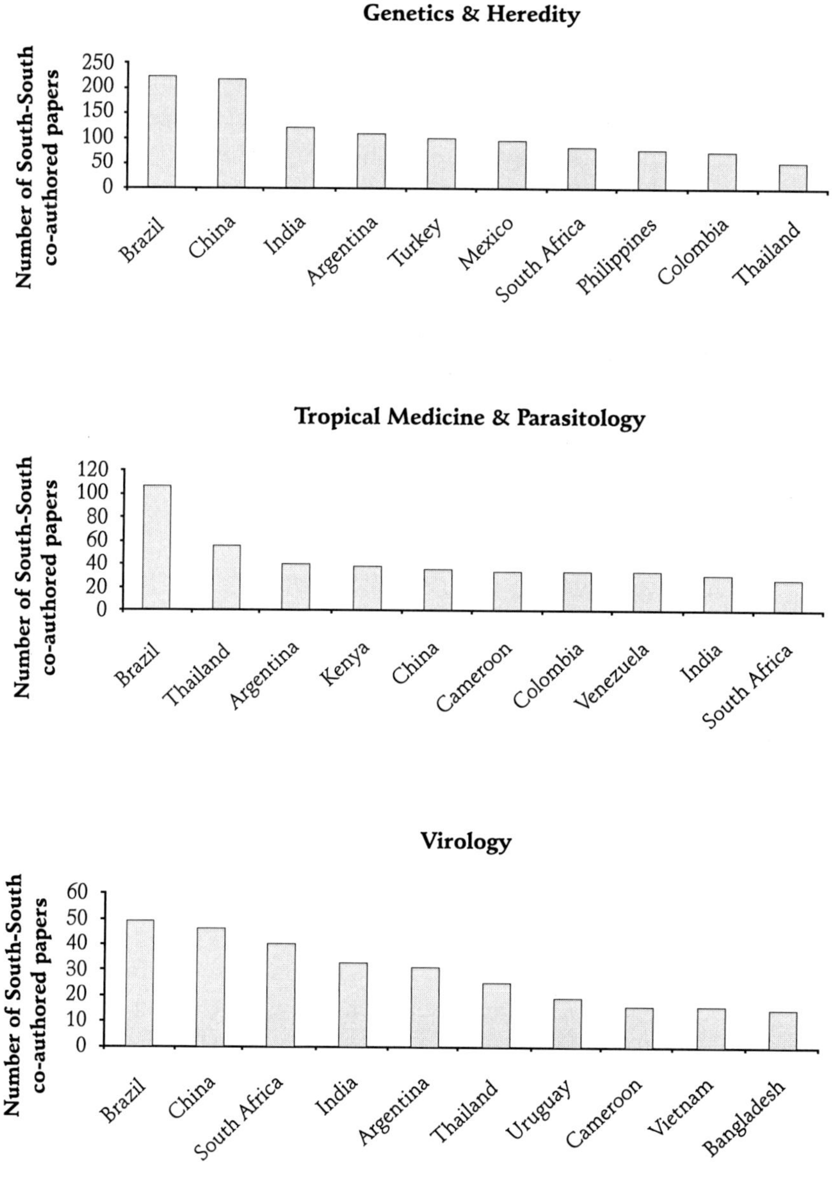

Source: Authors' presentation of data compiled by Science-Metrix using Scopus.

can either reflect genetic research on a common lineage, or it can be an indication that some developing countries have populations which are valuable for research in genomics, but that capacity is needed to harness this resource—something the leading developing countries in health biotechnology can provide. The South-South collaboration in Virology is also dominated by the leading developing countries in the field: Brazil, China, South Africa, and India in addition to Argentina.

In 'Tropical Medicine and Parasitology,' the situation is different with countries such as Kenya, Columbia, Venezuela, and Cameroon—generally weaker in health biotechnology—taking the lead with Brazil and Argentina. Therefore, depending on the subfield of health biotechnology, we observe diverse patterns of collaboration. Brazil is an exception as it is one of the main collaborators in all the subfields on which we focussed. It collaborates heavily in research on the tropical disease, Chagas disease, and almost a quarter of its collaboration with Argentina is focussed on research into this disease.

2.4.2 Increasing role of Africa in HIV/AIDS and malaria research collaboration

We looked further at South-South collaboration focussed on two diseases of special relevance to many developing countries, HIV/AIDS and malaria. When we looked at collaboration on HIV/AIDS research, we observed a somewhat different pattern in South-South linkages than in health biotechnology collaboration generally (Figure 2.8). Africa has become a more active collaborator around HIV/AIDS than more generally in health biotechnology. South Africa, for example, has frequent collaboration with Asian countries, particularly Thailand and with Brazil. It also collaborates with several sub-Saharan countries in HIV/AIDS research. Cameroon has also relatively active collaboration with China. In Asia, Thailand seems to be the most active country involved in South-South HIV/AIDS research. Apart from collaborating with South Africa, it collaborates frequently with Uganda and Brazil.

Argentina is another country that features rather prominently in South-South research on HIV/AIDS. It collaborates both with far away countries in Africa (particularly South Africa) and Asia (particularly Thailand) and

FIGURE 2.8

South-South collaboration network in HIV/AIDS research, 1996–2009

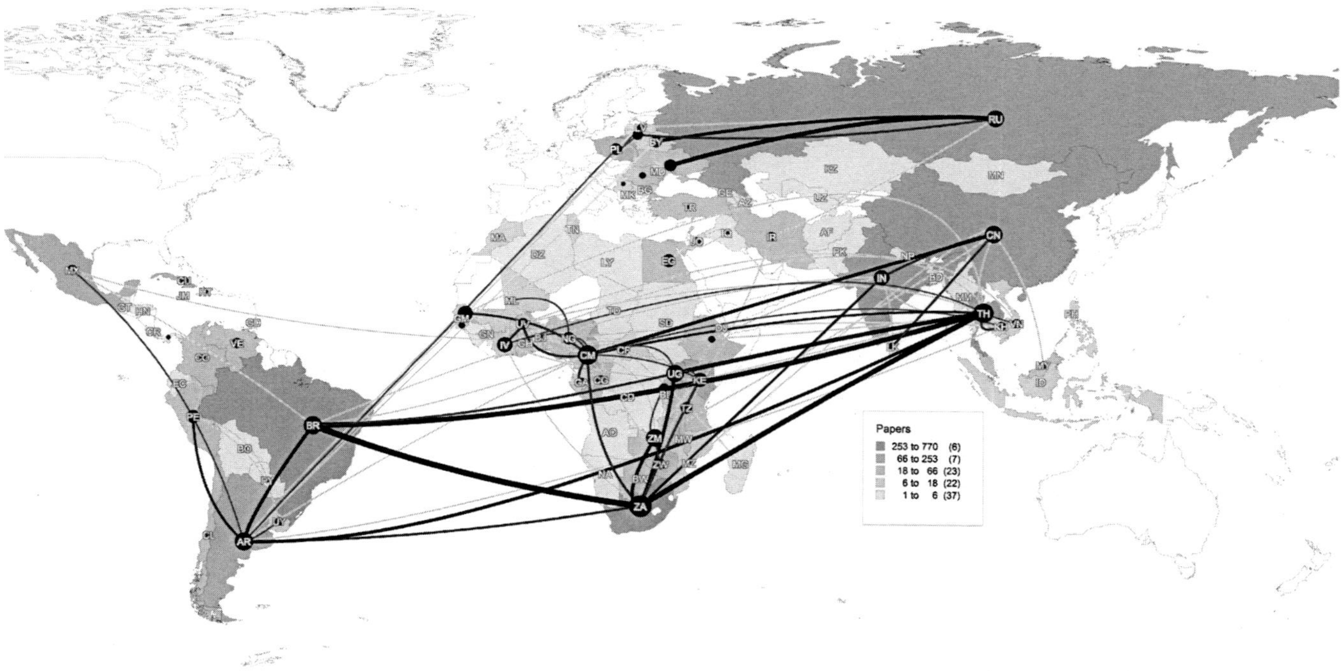

Note: The width of the lines represents the number of collaborations between two linked countries. To ease the representation of partnerships, only linkages of two or more co-publications were included on the map.

Source: Compiled by Science-Metrix using Scopus.

FIGURE 2.9

South-South collaboration network in malaria research, 1996–2009

Note: The width of the lines represents the number of collaborations between two linked countries. To ease the representation of partnerships, only linkages of two or more co-publications were included on the map.

Source: Compiled by Science-Metrix using Scopus.

also regionally with countries such as Peru and Uruguay. In HIV/AIDS research, there is also a separate cluster in the former Eastern Bloc countries where Russia collaborates particularly frequently with Ukraine and Belarus. It is also noteworthy that India does not seem to be actively collaborating with other countries on HIV/AIDS research and only has notable ties with China and South Africa. Further, even though Africa is relatively active in South-South HIV/AIDS collaboration, North Africa does not seem to be a member of that network.

When we look at South-South collaboration in malaria research we also see much stronger participation of Africa (Figure 2.9). What is noticeable in the malaria South-South collaboration involving Africa is that the primacy of South Africa is no longer evident. There seems to be active regional collaboration on malaria research involving relatively many African countries. Africa has for 15 years had formal networks in malaria research. The African Malaria Network Trust (AMANET) was established in 2002 but traces its origins in the African Malaria Vaccine Testing Network (AMVTN), which was established in 1995. These networks support infrastructure development and organise training in malaria surveillance and research as well as support research collaboration on malaria prevention and treatment. In the active African collaboration on malaria we may be observing impacts of these networks on malaria, a shared health problem. The strongest ties in African malaria research appear to be between Zimbabwe and Sudan, but we also observe frequent ties within Western Africa, for example, between Senegal and Ghana. In Asia, Thailand is particularly active in malaria research and collaborates particularly strongly with Burkina Faso, but also with Nigeria and Peru. We therefore observe active intercontinental collaboration that is unusual in the sense of not strongly involving the emerging economies, China, India and South Africa. Brazil appears as the only emerging economy relatively active in South-South malaria research. The patterns of South-South collaboration thus clearly differ depending on the focus of health biotechnology research.

By examining subfields we can identify two forces that seem to encourage South-South collaboration in health biotechnology. On the one hand, there are the forces that have encouraged international genomics research in general, fuelled by the belief that in order to tackle the complex challenge of genomics and to accelerate biomedical research, a global view of genomics is required. In addition there is a push towards working together on common health problems, where sharing expertise and resources can be a means of strengthening research. Further case-study research is needed to shed light on the importance of these different incentives in fuelling South-South research collaboration, and the different roles of developing countries in collaboration.

2.4.3 South-South research collaboration increases the visibility and impacts of developing countries' research

It is well known from previous research that papers published by researchers from developing countries are likely to have less global impact than papers from developed countries' researchers, partly due to the lower rate at which the former group publishes in high-impact journals (Arunachalam and Manorama, 1988; King, 2004; Osareh and Wilson, 1997). It is further known that international collaboration is likely to increase the visibility and impact of research in general, and is more likely to result in publications in high-impact journals than research without such collaboration (Glänzel and Schubert, 2001; Katz and Hicks, 1997). Therefore, it is of relevance to examine impact factors of co-authored South-South papers, and to compare them with the impact factors of developing countries' health biotechnology papers in general. We calculated the ARIF[2] of the developing countries' health biotechnology papers. This indicator reflects the citations a particular journal received relative to the number of papers it published. The impact factor of papers is calculated by ascribing to them the impact factors of the journals in which they are published. To adjust for different citation patterns across fields and subfields of health biotechnology, each paper's impact factor is divided by the average impact factor of the papers in its subfield. When the ARIF is above 1, it means that an entity scores better than the world average; when it is below 1, this means that on average, an entity publishes in journals that are not cited as often as the world average.

2. See definition of ARIF in Chapter 1.

The ARIFs of South-South co-authored papers are higher than the ARIFs of all developing countries' papers (Figure 2.10). Papers that developing countries' researchers publish in South-South collaboration are, therefore, seemingly more likely to be published in higher impact journals than papers they publish in general.

FIGURE 2.10

Comparison of ARIF of developing countries' health biotechnology papers in South-South collaboration versus all health biotechnology papers published by developing countries, 1996–2009

Source: Authors' presentation of data compiled by Science-Metrix using Scopus.

To further explore the potential impacts of developing countries' collaboration, we examined the impact factors in different subfields of health biotechnology. Table 2.2 shows that the subfields of health biotechnology that have most frequent South-South collaboration, have ARIFs approaching 1, the world level of relative impact factors for health biotechnology. The lowest value is in 'Biochemistry and Molecular Biology' (0.70) and the highest is in 'Tropical Medicine' (1.21). The latter field heavily includes contributions from developing countries, and is likely to be infrequently cited by developed countries' researchers. In 'Tropical Medicine,' developing countries' authors publish 58 per cent of all the papers published in the world.

TABLE 2.2

ARIF of health biotechnology papers in South-South collaboration for different subfields of health biotechnology, 1996–2009

Subfield of health biotech	ARIF
Genetics & heredity	0.89
Biochemistry & molecular biology	0.70
Biomedical engineering	0.83
Microbiology	0.90
Immunology	0.85
Virology	1.02
Tropical medicine	1.21
General biomedical research	0.94
Parasitology	1.03
Cancer	1.09

Source: Authors' presentation of data compiled by Science-Metrix using Scopus.

This analysis of health biotechnology (ARIFs) supports the notion that South-South collaboration is likely to increase the visibility and impact of developing countries' research. Still, according to our analysis, South-South co-authored papers have not reached the level of attention given to developed countries' papers in some of the subfields of health biotechnology.

2.5 Conclusions

The analysis above suggests that South-South research collaboration in health biotechnology has started to become a reality for developing countries. By examining South-South co-authorships of health biotechnology papers as a proxy for research collaboration, we observe that developing countries are increasingly engaging in research collaboration with each other. The levels of collaboration are still low and there is ample scope for increases.

Measuring collaboration through co-authorship levels is not without its limitations. It can both miss important collaboration that does not result in co-authored papers, and bolster collaboration when researchers are placed as authors on papers to which they have provided minimal contributions. Contacting every health biotechnology research group in a developing country and asking them about their South-South collaboration is not feasible. In the absence of such a massive survey, co-authorship can act as a proxy for collaboration, capable of identifying salient characteristics of the collaboration and showing changes over time that can then be linked back to policy initiatives. As such, this analysis serves as a baseline examination of South-South collaboration, to which future studies can compare.

In summary, our findings indicate that South-South collaboration in health biotechnology is:

Under-harnessed, but change may be on the horizon. The number of South-South co-authored papers published is generally low in health biotechnology when we look at all low- and middle-income countries and has not increased overall in the last decade. Considering that South-South collaboration is frequently a topic on the political agenda, and has clear benefits for health biotechnology research, it is somewhat surprising how few co-authored papers exist. However, there are some signs of change. Firstly, in the last few years a number of developing countries have made formal agreements to collaborate in science and technology, and a few have singled out health research and/or biotechnology as fields for the collaboration. The next decade will reveal if more South-South collaboration in health biotechnology will result from these agreements.

Secondly, the last few years have shown a relatively steep increase in South-South co-authored papers for the emerging economies of Brazil, China and India. Time will tell if this trend continues, and whether these countries will take on a larger role in promoting capacity building and contribute to innovation in health biotechnology amongst developing nations. Considering that these countries are experiencing high economic growth, this larger role would fit their standing amongst nations. The relatively recent increase in South-South collaboration among the leading developing countries, and the pattern of South-South collaboration involving international organisations, may suggest the need to overcome resource constraints and reach a certain threshold of resources in order to fully reap the benefits of collaborating with each other.

Successfully promoted by governmental thrust. The chapter also discussed Brazil's leading role in promoting South-South health biotechnology collaboration. It is the most active country in terms of South-South co-authored papers, and has had a policy thrust to promote collaboration with developing countries since 1985. The collaboration with Argentina has been particularly active and appears to be directly related to governmental policies promoting joint biotechnology development between the two countries, as well as the availability of funds to support collaboration. This suggests that when governments actively promote South-South collaboration, observable results can include an increase in co-authored papers published in international peer-reviewed journals.

Advanced by global forces promoting genomics. Our work here has also shown that South-South collaboration in health biotechnology is particularly strong in the subfield of 'Genetics and Heredity.' This resonates with the collaboration between high-income countries that has also been relatively strong in these areas. The emphasis on 'Genetics and Heredity' indicates that South-South collaboration is heavily influenced by global forces that promote international collaboration, such as the Human Genome Project. The South-South collaboration in 'Genetics and Heredity,' however, may not only be fuelled by efforts from developed countries, but also by a shared genetic lineage between the populations of developing countries, and the need for countries that lack genetics/genomics capacity to collaborate with their neighbours in order to harness genomics resources.

Shows African countries to be relatively active in HIV/AIDS and malaria research collaboration. Generally in health biotechnology, Africans do not seem to have membership in most Southern networks in the health biotechnology field. When we focussed particularly on South-South collaboration involving diseases of developing countries, HIV/AIDS, and malaria, we see a much stronger involvement of African countries and more active participation of other countries not strong in health biotechnology. Further research is needed to know if membership in these networks opens up access to knowledge and innovation potentials in this field or whether their participation is confined to be providers of samples and other local information.

Influencing international visibility of developing countries' research. This scientometric analysis also indicates that South-South co-authored papers are published in higher impact journals than papers generally published by researchers in developing countries; therefore leading us to believe that South-South collaboration may increase the impact of developing countries' research. Still, South-South research in many subfields of health biotechnology is generally published in journals that have lower-than-average citation rates in the international peer-reviewed literature. Engaging in South-South collaboration appears to help increase the impact of research by developing countries, but for several subfields it is still lower than the impact levels reached by health biotechnology research in general.

There are many questions about collaboration that can only be answered superficially by a scientometric analysis. For example, the analysis needs further clarification on what are the motivations for South-South collaboration in the health biotechnology field, what roles the different countries play in the collaborations, what challenges the collaborations impose, what impacts the collaborations have had, and how the collaborations can be strengthened. The next chapters address these questions and provide a fuller understanding through case-study research on how developing countries can leverage South-South collaboration to promote health research and innovation.

References

Arunachalam, S. and K. Manorama (1988). "How do journals on the periphery compare with mainstream scientific journals?", *Scientometrics* 14(1-2): 83-95.

Bozeman, B. and E. Corley (2004). "Scientists' collaboration strategies: Implications for scientific and technical human capital", *Research Policy* 33(4): 599-616.

Dandara, C., P.T. Basvi, T.E. Bapiro, J. Sayi and J.A. Hasler (2004). "Frequency of-163 C> A and 63 C> G single nucleotide polymorphism of cytochrome P450 1A2 in two African populations", *Clinical Chemistry and Laboratory Medicine* 42(8): 939-941.

Faulkner, W., J. Senker and L. Velho (1995). *Knowledge frontiers: Public sector research and industrial innovation in biotechnology, engineering ceramics and parallel computing.* Clarendon.

Glänzel, W. and A. Schubert (2001). "Double effort = Double impact? A critical view at international co-authorship in chemistry", *Scientometrics* 50(2): 199-214.

Hardy, B.-J., B. Seguin, F. Goodsaid, G. Jimenez-Sanchez, P.A. Singer and A.S. Daar (2008). "The next steps for genomic medicine: Challenges and opportunities for the developing world", *Nature Review Genetics*, Special supplement: Genomics medicine in developing countries. October.

Hassan, M.H. (2007a). "Building capacity in the life sciences in the developing world", *Cell* 131(3): 433-436.

———. (2007b). "A New Dawn for Science in Africa", *Science* 316(5833): 1813.

Katz, J.S. and D. Hicks (1997). "How much is a collaboration worth? A calibrated bibliometric model", *Scientometrics* 40(3): 541-554.

Katz, J.S. and B.R. Martin (1997). "What is research collaboration?", *Research Policy* 26(1): 1-18.

King, D.A. (2004). "The scientific impact of nations", *Nature* 430 (6997): 311-316.

Kong, Q.P., Y.G. Yao, M. Liu, S.P. Shen, C. Chen, C.L. Zhu, M.G. Palanichamy and Y.P. Zhang (2003). "Mitochondrial DNA sequence polymorphisms of five ethnic populations from northern China", *Human Genetics* 113(5): 391-405.

Osareh, F. and C. S. Wilson (1997). "Third world countries (TWC) research publications by disciplines: A country-by-country citation analysis", *Scientometrics* 39(3): 253-266.

Powell, W.W. and J. Owen-Smith (1998). "Universities and the market for intellectual property in the life sciences", *Journal of Policy Analysis and Management* 17(2): 253-277.

Rath, A. and S. Lealess (2000). *The forum on South-South cooperation in science and technology.* Seoul: Policy Research International Inc.

Thorsteinsdóttir, H., A.S. Daar, P.A. Singer and E. Archambault (2006). "Health biotechnology publishing takes-off in developing countries", *International Journal of Biotechnology* 8(1): 23-42.

Zucker, L.G., M.R. Darby and M.B. Brewer (1998). "Intellectual human capital and the birth of the US bioechnology enterprises", *The American Economic Review* 88(1): 290-306.

3 Mapping South-South Entrepreneurial Collaboration in Health Biotechnology: Boosting Trade and Innovation?*

AUTHORS: *Halla Thorsteinsdóttir, Christina C. Melon, Monali Ray, Sharon Chakkalackal, Michelle Li, Jan E. Cooper, Jennifer Chadder, Tirso W. Sáenz, Maria Carlota de Souza Paula, Wen Ke, Lexuan Li, Magdy A. Madkour, Sahar Aly, Nefertiti El-Nikhely, Sachin Chaturvedi, Victor Konde, Abdallah S. Daar, Peter A. Singer*

3.1 Introduction

So far the South-South collaboration we have mapped and discussed in Chapter 2, has been mainly between research groups at universities and public research institutions. It has been based on an analysis of co-authored papers listed in the Scopus database. As firms generally do not emphasise publishing in international peer-reviewed journals, an analysis of co-authorship is not likely to encompass firms' South-South collaboration. Developing countries' firms may also be working together in South-South partnerships with each other and we need to identify and apply a methodology that can measure their linkages.

* This chapter is a slightly revised version of a feature article published by the journal Nature Biotechnology (2010). "Entrepreneurial collaboration in health biotech", 28: 407–416. It is published with the permission of the journal.

Entrepreneurial firms in developing countries seem to be increasingly aware of each other's conditions and have begun targeting one another's markets, as can be observed in the increase in South-South trade at a rate of 12.5 per cent a year (OECD, 2006). This has partly been made possible by the changing landscape of development where emerging economies, such as China and India, have experienced unprecedented growth and increased global trade (The World Bank, 2010). Furthermore, as we have discussed in Chapter 1, developing countries have been setting up mechanisms to encourage increased trade with one another, by establishing free trade zones such as the ASEAN Free Trade Area, the Southern Common Market (Mercosur/Mercosul) in Latin America, and the Common Market for Eastern and Southern Africa (COMESA).

In the health fields analysts have made calls for increased South-South collaboration to address shared health problems (Morel *et al.,* 2005). Developing countries are increasingly aware of the importance of doing so through joint efforts with one another, and have set up networks to deal with malaria, tuberculosis (TB), HIV/AIDS, and other common diseases. Brazil, China, Cuba, Nigeria, Russia, Thailand and Ukraine are working together in a network that jointly promotes R&D aimed at developing innovative diagnostics kits, drugs, and vaccines for HIV/AIDS prevention and treatment (Lemle, 2005). Also 24 manufacturers of vaccines in developing countries form the Developing Countries Vaccine Manufacturers Network (DCVMN) to ensure a consistent and sustainable supply of quality vaccines at an affordable price to developing countries and to encourage R&D efforts to meet the emerging vaccine needs in the developing world (*www.dcvmn.com*).

There is, however, a limited amount of empirical evidence on the collaborations amongst firms in developing countries. In health biotechnology, for example, no research confirms that companies in developing countries' have heeded the call for South-South collaboration, or that they are to any significant degree working together. The goal of this chapter is to help fill this gap and provide empirical knowledge on collaboration between health biotechnology firms in developing countries, what we call South-South firm collaboration. Collaboration between firms in developing and developed countries, is defined as South-North firm collaboration in the context of this research.

3.2 Rationale for South-South collaboration

One reason why firms in health biotechnology, both in developing countries and elsewhere, may want to work together is the need to minimise costs and risk. The development of new health products and services in biotechnology are characterised by high costs and high risks and many lead candidates fail after costly clinical trials have been carried out (Pisano, 2006). Working in collaboration also appeals to firms as it is a way to gain access to new markets. Alliances between firms in different countries can facilitate their market access and help them expand their markets (Hagedoorn, 2002).

Firms in small countries are dependent on exporting their products in order to survive. It can be appealing for them to collaborate with firms in other countries in order to obtain this access. Further access to strategic knowledge, and particular technical skills is an important reason for collaboration between firms (Faulkner *et al.,* 1995; Hagedoorn, 2002; Lee, 2007; Pisano, 2006; Roijakkers and Hagedoorn, 2006). The knowledge required for product development in health biotechnology is highly specialised, making it nearly impossible for small firms or institutions in developing countries to possess the spectra of knowledge needed. Collaborations therefore become a tool for firms to obtain access to a wide spectrum of knowledge, technologies and skills. The knowledge can be requisite for various phases of health biotechnology development and for instance, access to regulatory knowledge is key to their innovation potential.

If developing countries can cultivate ways to work effectively together, they may be able to harness a more relevant model of promoting innovation than the traditional model of relying on linkages with developed countries. They could strengthen their capability to address shared problems by pooling their expertise and resources—problems that developed countries may not be affected by, nor be interested in. If successful, South-South collaboration could increase capacity in science-intensive fields by learning from each other, improve the ability of developing countries to address their own problems, and contribute to economic development and quality of life in developing countries.

To examine the level and characteristics of South-South collaboration, we sent a brief survey to 467 health biotechnology firms in six developing countries—Brazil, China, Cuba, Egypt, India and South Africa—and asked about their linkages with all other developing countries (see further details of the methodology in Chapter 1). We chose countries that our previous research has identified as being relatively strong in health biotechnology or regional leaders in the field (Thorsteinsdóttir *et al.,* 2004a). We sent the survey to all dedicated biotechnology firms in these countries we could identify, pharmaceutical firms active in biotechnology, as well as other organisations that are heavily involved in commercialisation activities in the health biotechnology field. We asked the firms whether they collaborated with firms/organisations in other low- and middle-income countries, and if so, to name their collaborators and provide an overview of each partnerships. Data collected included the reasons for the collaboration, the activities involved, and the output of the collaboration. We presented the firms with a broad definition of 'collaboration', and included in that definition any work jointly undertaken by firms and organisations which contributes to the production of knowledge, products, and services in health biotechnology.

A total of 288 firms completed the survey resulting in a response rate of 62 per cent (Table 3.1). We feel this is a solid response rate, given that participation was voluntary and the nature of the sector can make it challenging to get responses from firms. The sector is fluid with companies

TABLE 3.1

Number of health biotechnology firms surveyed and their response rates

Country	Number of firms surveyed	Number of responses	Response rate (Per cent)
Brazil	110	72	66
China	139	83	60
Cuba	11	8	73
Egypt	22	15	68
India	121	68	56
South Africa	64	42	66
Total	467	288	62

Source: Authors' presentation of their own survey data.

frequently merging and/or going bankrupt. In OECD biotechnology surveys involving mandatory responses, only response rates under 50 per cent are considered to be low (van Beuzekom and Arundel, 2006).

In this chapter, we present the results of the extent of the South-South health biotechnology collaborations, map where the main linkages lie, and explore the main characteristics and outputs of the collaborations.

3.3 Extent of South-South collaboration

The results show that South-South firm collaboration is significant, with more than a quarter (27 per cent) of the health biotechnology firms that responded reporting collaborations of this type (Figure 3.1). South-North collaboration is still more predominant, with over half (53 per cent) of the firms reporting collaborations with developed countries. Many of the firms (21 per cent) indicated both South-South and South-North collaborations.

FIGURE 3.1

Examination of the extent of international collaboration of health biotechnology firms in developing countries and comparisons of their South-South versus South-North collaborations

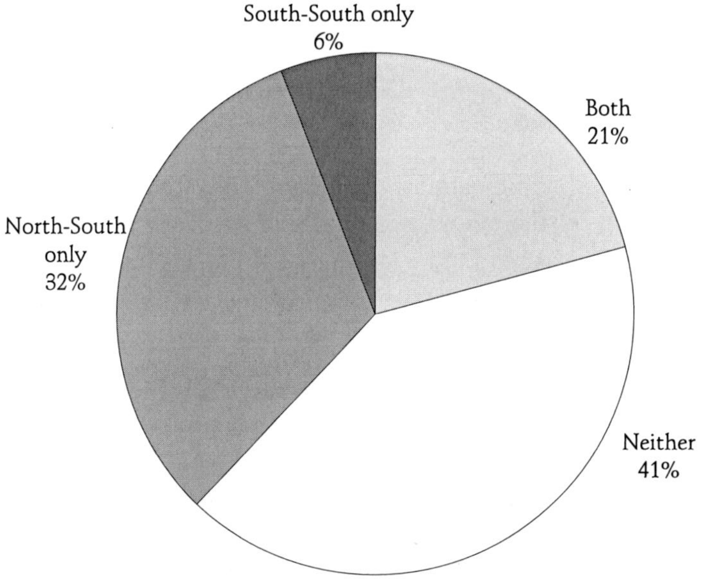

Source: Authors' presentation of their own survey data.

We looked at the proportion of firms involved in South-South collaboration in each of the countries we studied (Figure 3.2). Those with the smallest populations—Cuba and South Africa—are the most active in South-South collaborations, with almost half of the South African firms and three quarters of the Cuban entrepreneurial organisations reporting involvement in this type of collaboration. This is in stark contrast to the more populated countries, such as China, where just over 10 per cent of the firms report South-South collaborations, and India with less than 20 per cent of firms doing so.

FIGURE 3.2

Percentages of firms in the countries we surveyed that engage in South-South health biotechnology collaboration

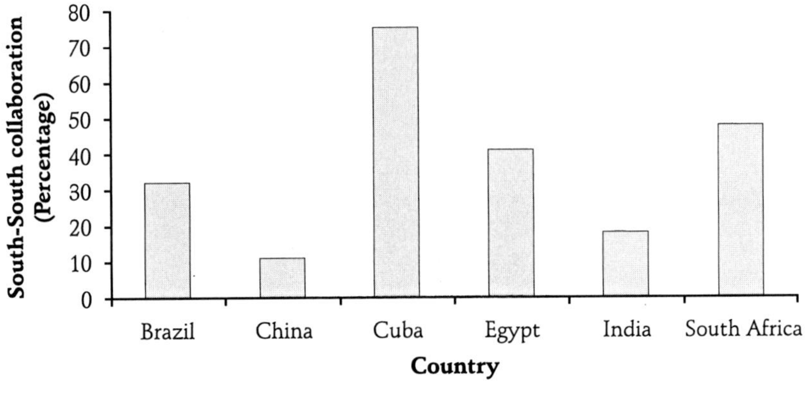

Source: Authors' presentation of their own survey data.

According to our findings, almost all the countries studied are more active in South-North collaborations than South-South collaboration. Egypt was the only country that showed a lower rate of South-North collaboration, with twice as many South-South collaborations as South-North (Table 3.2).

Most of the firms that are active in South-South collaboration are engaged in several collaboration initiatives. The total number of South-South collaborations reported in this study was 279. However it is important to note that some collaborations may have been double-counted, i.e., a particular collaboration between India and South Africa may have been counted twice—once for India and once for South Africa—if both firms responded to the survey and reported all of their collaborations. To address this issue, we asked the respondents to provide the names of their partnering firms, but many opted to keep this information confidential, thereby limiting our ability to adjust the number of collaborations accordingly. In such cases, the firms reported, for example, that they collaborated with Firm A in India and Firm B in China. This may inflate the number of South-South collaborations when presented in an aggregate form.

TABLE 3.2

Number of international collaborations reported

Country	South-South collaboration		South-North collaboration		Total collaboration	
	Number	Average number per company	Number	Average number per company	Number	Average number per company
Brazil	64	0.9	127	1.8	191	2.7
China	27	0.3	99	1.2	126	1.5
Cuba	34	4.3	63	7.9	97	12.1
Egypt	39	2.6	30	2.0	69	4.6
India	54	0.8	126	1.9	180	2.6
South Africa	61	1.5	66	1.6	127	3.0
Total	279	1.0	511	1.8	790	2.7

Source: Authors' presentation of their own survey data.

Brazil has the largest number of South-South collaborations of the countries we surveyed, with well over 60 collaborations. Even though the countries with the smallest populations, Cuba and South Africa, have a relatively low number of health biotechnology firms, they are so active in South-South collaborations that comparing their collaborations with those of large countries is still likely to produce valid results. South Africa has the second highest number of collaborations of the countries in this study, and Cuba has slightly more collaborations than the population giant, China.

We asked the firms to indicate who initiated the collaborations—themselves, their partners, government agencies, international organisations, expatriates, or any other intermediary. Their answers indicate that the firms themselves typically initiated the collaboration. Governments or other local or international organisations seldom played this role, with only 17 of the 279 reported collaborations said to have been initiated by these organisations. Respondents from Cuba and Brazil were most likely to indicate governmental roles and, typically, these involved their public research organisations that are heavily involved in entrepreneurial activities. As can be seen in the later chapters of this book, interview evidence confirms that firms find it challenging to identify appropriate partners in other developing countries to collaborate with and initiate the collaboration. Finding enough detailed information about their potential partners is a difficult task and trust building can be challenging. There is thus a definite scope for the further role of governments and other third parties in initiating the collaboration. It is also notable that only one of the collaborations was reported to be initiated by expatriate residents of the collaborating countries. One explanation for this may be a relatively low migration rate of professionals between developing countries, and it would be interesting to see if expatriates play a larger role in South-North health biotechnology collaboration.

In addition, we asked the respondents to indicate if they had set up formal arrangements with their collaborators, and to elaborate on their nature where applicable. We found that most (almost 90 per cent) of the collaborations involved at least one type of formal arrangement among participants, ranging from supply agreements, to R&D cooperation agreements, to marketing and distribution agreements. Licensing agreements were commonly cited, with around 19 per cent of the collaborations having formal licensing contracts, while joint ventures were only established in around 8 per cent of the collaborations overall. South Africa (seven joint ventures), and Cuba (six joint ventures) had the highest numbers of joint ventures reported.

3.4 Geography of South-South entrepreneurial collaborations

To examine where the South-South collaborations in health biotechnology are, we drew a graph of the main linkages reported by the firms using the UCINET 6 programme. From Figure 3.3, we see that the countries of focus are all hubs involved in various collaboration networks. As these are the countries we surveyed directly, it is not surprising that they are featured centrally. Figure 3.3 is also likely to under-represent the collaborations

FIGURE 3.3

Collaboration network of health biotechnology firms in South-South collaborations

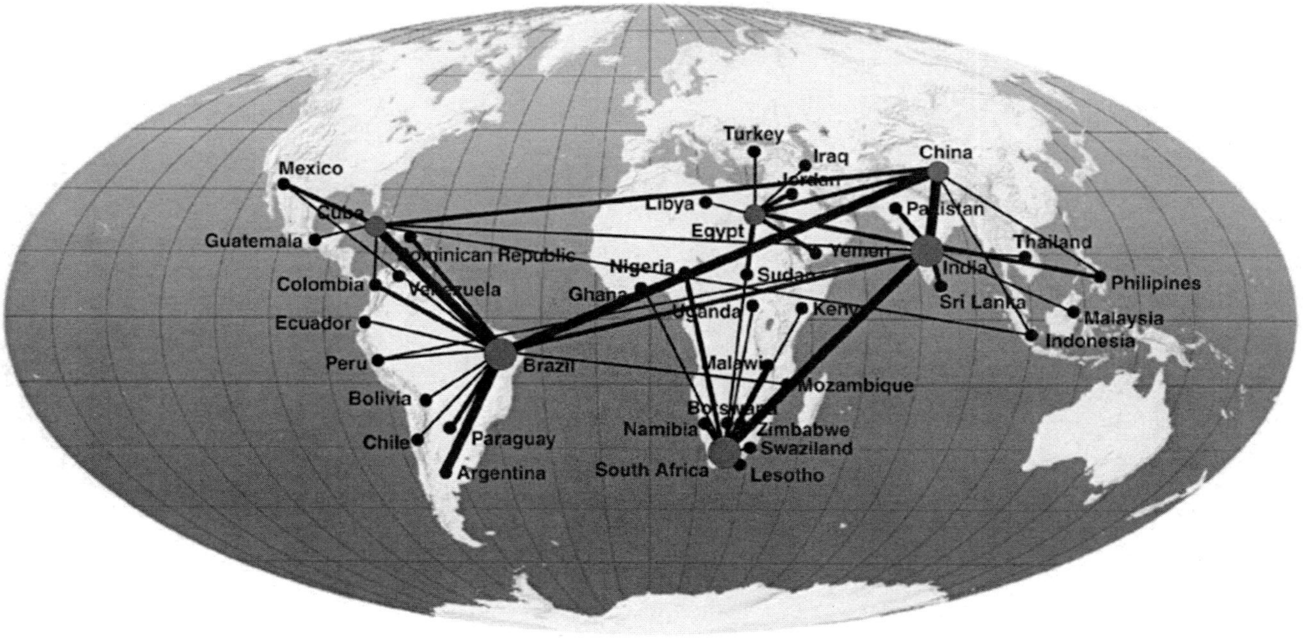

Note: The size of node represents the total number of South-South collaborations for the country, while the width of the lines represents the number of collaborations between the two linked countries. To ease the representation of partnerships, only linkages of two or more collaborations were included on this world map.

Source: Authors' presentation of their own survey data.

of countries we did not survey, such as Mexico, Nigeria and Malaysia. Nevertheless, the map provides a solid overview of the South-South collaboration and shows that the strongest linkages of the countries we surveyed are with one another. Chinese companies collaborate mainly with those in Brazil and India, Indian companies have close linkages with those in South Africa, and Brazilian companies have close linkages with entrepreneurial units in Cuba. The only other pair of countries where companies are involved in a similar level of South-South collaborations are Brazil and Argentina, and South Africa and Botswana. As Argentina and Botswana are active in South-South collaboration, surveying them would have provided an even fuller picture of South-South firm collaboration in this field. Our data, however, reinforce the notion that we surveyed the strongest countries in health biotechnology, and that they collaborate with one another despite substantial distances.

The mapping further shows that there are considerable regional collaborations in the health biotechnology field. All the countries in our survey have collaborations with other countries within their continent. It is, for example, notable how many ties South Africa has with other sub-Saharan countries; Egypt collaborates with Middle Eastern and North African countries; and there are considerable linkages between Brazil and Cuba with other Latin American countries.

3.5 Characteristics of South-South entrepreneurial collaborations

The results so far tell us that there are collaborations between developing countries' firms in health biotechnology. To get a deeper understanding of these collaborations, we asked the firms: what activities were involved in the collaborations, what were the reasons for their collaborations, and what outputs they had led to.

3.5.1 Collaborations involve mostly commercialisation

We asked the firms to specify the activities they were pursuing jointly in South-South collaboration. We provided them with a wide selection of activities that are typically undertaken by health biotechnology firms, from activities that are research intensive, i.e., R&D, to end-stage commercialisation activities such as distribution and marketing. We considered activities to be innovative if they focussed on research and any developmental activities of new products or services or of production processes. This includes, for instance, clinical trials and laboratory services. Conversely, we regarded collaboration involving simply the packaging of products or their export between countries—as non-innovative activities. We indicated to the firms that they should choose all the activities that were applicable to their collaborations, and offered the option to include any other activities not included on our list.

The results show that the majority of the collaborations (60 per cent) involve two or more activities. For example, rather than creating collaboration solely around distribution, collaborations are more likely to involve distribution and another activity, such as providing supplies. It is also clear that most of the South-South collaborations involve end-stage commercialisation activities, with around 200 (72 per cent) of the collaborations involving distribution, and 95 (34 per cent) involving marketing activities (Figure 3.4). Developmental activities were much less frequently cited by the firms that responded: R&D was part of only 36 (13 per cent) of the collaborations, clinical trials just 25 collaborations (9 per cent), and contract research only 9 collaborations (3 per cent).

FIGURE 3.4

Distributions of the activities involved in the South-South entrepreneurial collaborations for all the countries we surveyed

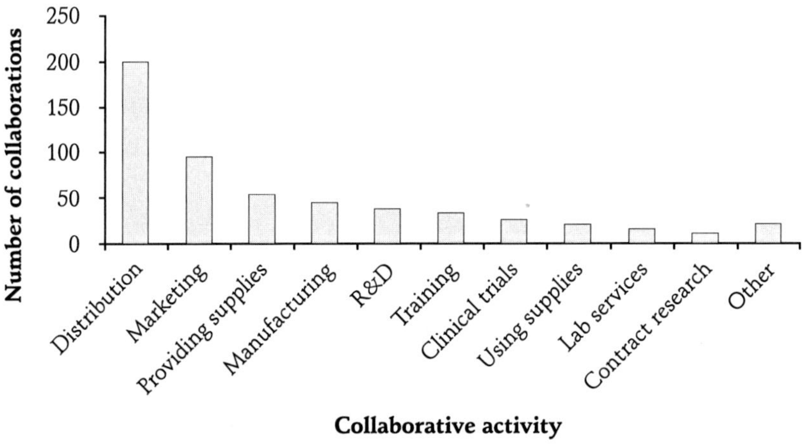

Source: Authors' presentation of their own survey data.

It is noteworthy that the third most frequently cited collaboration activity was providing supplies, with 53 (19 per cent) of the South-South collaborations involving such provisions. Potential supplies can vary from providing plant material for drug development to providing active pharmaceutical ingredients.

The relatively small emphasis on R&D activities in South-South firm collaboration reflects less emphasis on developmental activities than reported in an analysis of North-North collaborations in biotechnology (Roijakkers and Hagedoorn, 2006). From the mid-to-late 1990s, more than 20 per cent of biotechnology collaborations between developed countries involved R&D, up from around 6 per cent in the 1970s. It will be of interest to repeat this exercise in a few years to detect if R&D collaborations between developing countries also increase.

We then explored where the collaboration linkages lie for the different types of activities (Figure 3.5). Some of the activities represent only a few collaboration linkages which certainly limits the possibility of generalising from these results. As distribution and marketing are closely related activities, we graphed them together as 'end-stage commercialisation'. We can see in Figure 3.5(a) that there are relatively strong end-stage commercialisation linkages between the leading developing countries in health biotechnology, with, for example, active distribution and marketing collaborations between Brazil and China, Brazil and Cuba, India and China, and India and South Africa. They likely form linkages to reach each others' markets. Also striking is the widespread regional commercialisation collaborations in health biotechnology. South African firms for example, have distribution and marketing collaborations with well over 20 African countries, including relatively strong linkages with Botswana, Namibia and Nigeria. Egypt has distribution and marketing collaborations with around 10 African countries, and widely within the Middle East. India has commercialisation collaborations with other Asian countries, such as Sri Lanka and Pakistan. Brazil has a relatively large number of commercialisation collaborations with other Latin American countries, though it should be noted that its only commercialisation collaborations in Africa are with Portuguese-speaking countries such as Angola and Mozambique. According to our survey, Brazil and South Africa

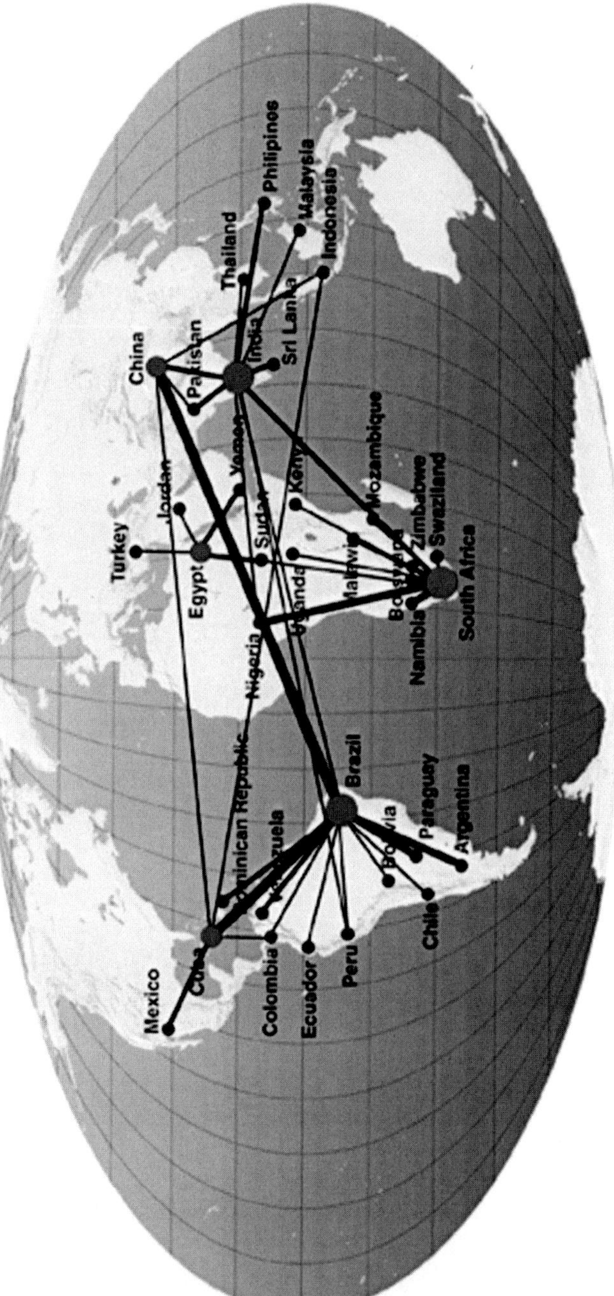

FIGURE 3.5
The network of collaborations involving end-stage commercialisation versus R&D

3.5(a). End-stage commercialisation

3.5(b). R&D

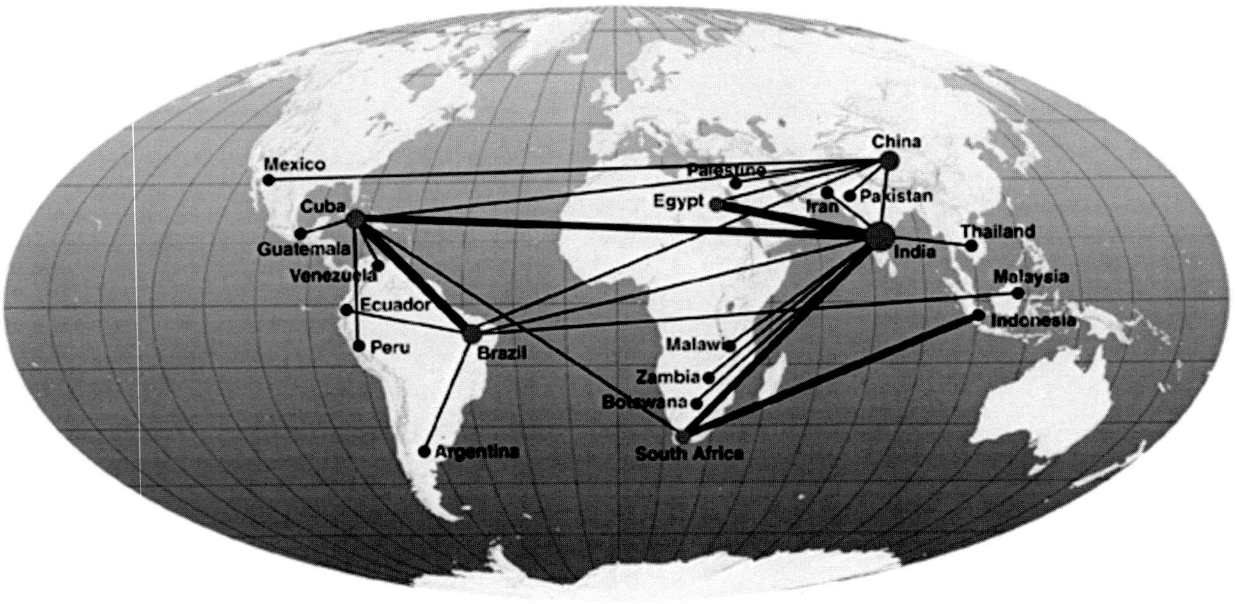

Note: The size of node represents the total number of South-South collaboration in the particular activity for a country, while the width of the lines represents the number of such collaborations between the two linked countries. To ease the representation of the distribution and marketing linkages, only countries which participated in two or more collaborations were included on this world map. For R&D, we present all of the linkages that were reported.

Source: Authors' presentation of their own survey data.

do not have distribution and marketing linkages in health biotechnology with each other, nor do Egypt and South Africa.

Further observations show that China has frequent collaborations with both India and Brazil in providing supplies. It is also notable that South Africa mainly provided supplies to other sub-Saharan countries. This may suggest that its collaborations are focussed on providing necessary products or ingredients for biotechnology development in countries with limited capacity in this field, including the provision of active pharmaceutical ingredients. The case-study research discussed in the later chapters of this book has supported this notion.

The survey data suggest that India and China are most active in manufacturing collaborations, which is not surprising as manufacturing in general is an area of strength for both countries (Bower and Sulej, 2007; Chaturvedi *et al.,* 2007; Yusuf *et al.,* 2007). Their manufacturing collaborations appear mainly to be intercontinental, between the leading developing countries, with relatively strong ties between China-Brazil, India-South Africa, and India-Egypt. The large markets in China and India can encourage smaller countries to set up joint ventures there around the manufacturing of innovations from the small countries. This is the preferred approach to exporting finished products from the small countries, which would result in high transportation costs.

3.5.2 The limited R&D collaborations are centred around a few countries

It is obvious from looking at Figure 3.5(b) that R&D collaborations are not nearly as active as end-stage commercialisation collaborations. The main linkages in R&D are between the leading developing countries in health biotechnology, the strongest of which are between Brazil and Cuba, India and Egypt, Cuba and India, and India and South Africa. Cuba and India seem to have relatively strong R&D collaborations compared to their end-stage commercial collaborations. There are however, other active R&D linkages between developing countries; South Africa and Indonesia for example, do joint R&D activities, and Cuba, India and China have low levels of R&D collaborations with several countries. In the case of Cuba, these are mostly regional collaborations with other countries in Latin

America, whereas India's collaborations are cross-continental and involve several African countries. Additionally it is notable that China and India seem to be more heavily involved in collaborations surrounding end-stage commercialisation and rarely collaborate in R&D activities with each other.

Developing countries conduct joint R&D for a variety of product development. Vaccines play a key health prevention role in developing countries and by working together they can strengthen their potential for developing cost-effective vaccines, targeting shared health problems in developing countries. Cholera is a shared health problem in Bangladesh and eastern India. The International Centre for Diarrhoeal Disease Research (Dhaka, Bangladesh) has been conducting leading research on cholera vaccines candidates, and as discussed in Chapter 7, by teaming up with the Indian firm Biological E (Hyderabad, India), they were able to start furthering the development of a cholera vaccine candidate. If successful they can eventually gear up towards manufacturing of the vaccine by the Indian firm. Another example of a vaccine R&D is discussed in Chapter 4 and involves the Bio-Manguinhos (Rio de Janeiro, Brazil), in collaboration with the Finlay Institute (Havana, Cuba). They utilised each others' respective strengths to develop and manufacture a bivalent meningitis AC vaccine to address a meningitis outbreak in Africa. This is a good example of how developing countries can use their assets in biotechnology to address health problems of other countries in need. Developing countries do not only work together in vaccine research but also on diverse health applications, including regenerative medicine. As discussed in Chapter 8, the South African firm Altis Biologics (Pretoria, South Africa) works, for example, with the First Affiliated Hospital of Xinjiang Medical University (Xinjiang, China) on morphogenic bone remodelling protein studies in rabbits to gather preclinical data.

Our survey results indicate that South-South collaborations rarely include clinical trials (another developmental activity). There are, however, some interesting cases of South-South collaboration involving clinical trials. Cuba seems to have the most active clinical trial collaborations of the countries we examined. Some of these collaborations involved South-South-North collaborations. CIMAB SA (Havana, Cuba), with its partner YM BioSciences Inc. (Mississauga, Canada), has spearheaded the establishment of a global clinical consortium to test cancer therapeutics based on original innovation from Cuba (Box 3.1). The network includes partners from 20 developing countries and thus, has a heavy emphasis on South-South collaboration. China is also involved in South-South collaboration focussed on clinical trials. As discussed in Chapter 5, the Chinese firm SH-IDEA Pharmaceutical Company (Yuxi, China) and the Kunming Institute of Botany (Kunming, China) are working with Thailand's Ministry of Public Health (Bangkok, Thailand) on clinical trials of HIV/AIDS treatment. The study stems from original research from the Kunming Institute of Botany based on Chinese traditional medicine and local biodiversity, but the clinical trials were carried out on Thai patients.

BOX 3.1

Global South-South-North consortium for clinical trials

To carry out cost-effective clinical trials, CIMAB SA (Havana, Cuba), the commercial arm of Cuba's Center of Molecular Immunology (Havana) and its partner YM BioSciences (Mississauga, Canada), have established a consortium of firms around the world for testing the humanised monoclonal antibody nimotuzumab in the treatment and diagnosis of patients with cancers of epithelial origin. The consortium has partners from 20 developing countries as well as seven developed countries. They include Argentina, Brazil, Columbia, Mexico, Peru, Paraguay and Uruguay from Latin America; Egypt, Morocco, Algeria, Nigeria and South Africa from Africa; and China, India, Pakistan, Indonesia, Malaysia and the Philippines from Asia. Asia is especially strong in the consortium, with Japan, Singapore and South Korea as developed country participants. Other high-income countries in the network are Saudi Arabia and Germany. The consortium thus reflects a South-South-North collaboration with strong participation from developing countries. Examples of Southern firms in the consortium are Biocon Biopharmaceuticals (Bangalore, India), Biotechnology Pharmaceutical Co. (Beijing, China), Eurofarma (Sao Paulo, Brazil), and Laboratorios PiSA (Guadalajara, Mexico).

Nimotuzumab is a Cuban innovation from the Center of Molecular Immunology that targets epidermal growth factor receptor. It is aimed at various epithelial cancer types, including non-small-cell lung, glioma, esophageal, brain metastasis, colorectal, pancreatic, prostate, cervical, and breast cancers. To date the consortium has tested nimotuzumab in

contd...

contd...

9,842 patients in Cuba, Argentina, Brazil, Canada, China, Colombia, Germany, India, Indonesia, Japan, Malaysia, Mexico, Singapore, South Africa, South Korea, Thailand and the Philippines. Trials are also being conducted in Europe, Japan and North America. CIMAB SA and YM BioSciences work to ensure that the network of firms follows the International Conference on Harmonisation of Technical Requirements for Registration of Pharmaceuticals for Human Use/Good Clinical Practice regulatory guidelines. The consortium's clinical trial results are collected in a central depository. Aggregating patient data from sites in the various countries increases the statistical power and quality of the clinical trials. By amassing data gathered under internationally recognised norms from the collaborating sites, the partners are able to submit a stronger drug application to their national regulatory authorities. Gaining approval from one regulatory agency can pave the way for other agencies to be able to approve the product. Currently nimotuzumab has been approved for marketing, as a treatment for head and neck cancers and glioma in 23 countries worldwide, including Argentina, Brazil, China, India, Indonesia, Mexico and Ukraine. The consortium members license the drug from CIMABSA and market it in their home countries.

Running clinical trials in developing countries among several partners has a number of advantages. Economies are obtained through the lower personnel and infrastructure costs and by sharing clinical trials expenses across several partners. Patient recruitment is faster, even for rare cancer indications, owing to the large populations who previously lacked access to treatments. Thus, not only are costs reduced, but trials are completed at a faster pace. The example of nimotuzumab shows that a consortium of enterprises, consisting primarily of small biotechnology firms from developing countries can complete these studies at the same speed as, and at lower cost than, big pharma. By including a South-South collaboration strategy, biotechnology firms can bypass pharma companies in clinical development and can potentially retain greater presence in later stages of the product's development and a greater share of revenue streams.

It should also be noted that, according to our survey, the South-South collaboration of Indian firms in clinical trials is limited. As India is known for active international collaborations involving clinical trials (Bower and Sulej, 2007; Chaturvedi *et al.,* 2007; Maiti and Raghavendra, 2007), its lack of clinical trial partnerships with other developing countries perhaps reflects the greater allure of relationships with multinational pharmaceutical firms or with developed countries.

3.5.3 Bi-directional knowledge flow is an important reason for the collaboration

To better understand the motivations for South-South firm collaboration, we asked the respondents to indicate the reasons for each of their collaborations. Again we note the multifaceted nature of South-South collaborations with respondents reporting several reasons for single collaborations. In line with the heavy emphasis on end-stage commercialisation collaborations, 'access to markets' was the main reason given for the collaborations (207 or 74 per cent of the collaborations). It was an important reason for the collaborations of all the countries we surveyed; firms in developing countries are clearly working together in order to gain export markets for their products/services. The second most commonly cited reason for the collaborations was to 'provide knowledge' (72 or 26 per cent of the collaborations), followed by to 'gain knowledge' (52 or 19 per cent of the collaborations). A relatively high proportion of Cuban respondents (68 per cent) cited 'provide knowledge' as a reason for the collaboration. Brazilians also cited this reason fairly often, but they, more frequently than the Cubans, reported knowledge gain as a reason for their collaborations.

There is mention of clinical access as a reason with 'access to patients' stated for 28 (10 per cent) of the collaborations, mainly by Chinese and Cuban respondents. Finally, 'provide patients' was a factor in 13 (5 per cent) of the reported collaborations. What is notable is how infrequently financial reasons were given for the collaborations with 'access to financing' cited as a reason for only 15 (5 per cent) of the collaborations, and 'providing financing' only cited four times (1 per cent). Cubans stood out again in citing 'access to financing' relatively frequently as a reason for their collaboration, as well as to 'provide technology/equipment'. This may suggest that they have collaborations that involve licensing access to their technologies to other developing countries.

It is noteworthy how relatively frequently 'provide knowledge' and 'gain knowledge' are cited as reasons for collaborations, especially given how rarely activities related to R&D were reported in our study. It points to a strong capacity-building role of the collaboration as we can see in examples of technology transfer initiatives. This may mean that the collaboration is still in its infancy, although its aim is future knowledge generation activities. This discrepancy may also reflect that different types of knowledge are required in health biotechnology. South-South collaboration may be used to gain access to knowledge about each others' markets, or to deal with regulatory affairs, etc.

Some of the reasons reported here align well with reasons attributed for the North-North or South-North collaborations (Faulkner *et al.,* 1995; Hagedoorn, 2002; Lee, 2007; Pisano, 2006; Ray *et al.,* 2009; Roijakkers and Hagedoorn, 2006). Access to markets and knowledge are both consistent incentives. Still, compared to findings from developed countries, where the need to access financing and minimising costs regularly stimulates collaboration, we would have expected access to financing to be cited more often as a reason for the South-South collaborations than it was. Based on the survey results, therefore, we cannot conclude that the South-South collaborations were fuelled by motivations to minimise costs.

3.5.4 The collaborations are strongly product focussed

We asked the respondents of the survey to report the outputs of their South-South collaboration. The majority of collaborations, roughly 65 per cent, have resulted in some specific output. The collaborations are strongly product focussed, with 70 collaborations (25 per cent) leading to a joint product in the market and 16 collaborations (6 per cent) leading to a joint product in the pipeline. South-South collaborations thus seem to facilitate the end-stage commercialisation of health biotechnology products produced by firms in developing countries, and increase their availability in developing countries. However, it is not likely that the South-South collaborations will result in the development of joint products, but instead will be confined to licensing arrangements. Only 16 collaborations (6 per cent) led to joint products in the pipeline, and joint patents were only reported as an outcome for 12 (4 per cent) of the collaborations. Cuban and Brazilian enterprises were the only ones that reported joint patenting as an outcome of their collaborations. Not surprisingly, South-South firm collaboration seems to rarely result in joint publications, and it was only reported once as an output from collaboration. Other reported outputs included: clinical/scientific research results, human resource training, separate product development, and technology transfers.

Our analysis also reveals that more than half of the collaborations involving R&D had joint products on the market, and a quarter of them had joint products in the pipeline. Even though there is generally a limited emphasis on product development in the South-South collaborations examined here, we can see that product development and end-stage commercialisation activities are closely linked. Now, a number of developing countries are signatories of the TRIPS agreement and their firms have started to place an increasing emphasis on R&D and developing 'new to the world' innovation (Chaturvedi, 2007; Frew *et al.,* 2007; Kale and Little, 2007; Simonetti and Archambault, 2007). Our survey results suggest that those firms may, in part, be relying on their commercialisation linkages with other developing countries to jointly strengthen their R&D activities. This is a promising sign that South-South collaborations will, in the future, play an important role in strengthening health biotechnology innovation within developing countries.

3.6 Conclusions

Our analysis indicates that South-South entrepreneurial collaboration in health biotechnology is significant and firms in developing countries are actively working together. It is, however, stronger in boosting trade rather than innovation in the health biotechnology field seen in the development of new products or processes. Apart from providing insight into the level and characteristics of South-South collaboration, our survey also establishes a baseline to which future studies can compare. As such, it can provide important information for evaluating the effects of South-South collaboration policies and programmes. As with any survey, we cannot claim our study is without its limitations. For logistical reasons, we had to limit our data collection to a few select countries: those that are likely to represent the bulk of developing countries' firms active in this field. Furthermore, we have not been able to receive information from every firm active in health biotechnology in the countries we focussed on, and some firms may not have reported to us the extent and characteristics of all their South-South collaborations. Still, as we obtained a relatively high response rate, we believe that the results represent the main characteristics of South-South firm collaboration in the health biotechnology field.

In summary, our findings lead us to draw several conclusions on South-South firm collaboration in health biotechnology. They include the following:

South-South collaboration has become a widely taken route for health biotechnology firms. The results of our study showed that one in every four firms that responded to our survey stated an active collaboration with other developing countries. Further, developing countries' firms that engage in South-South collaboration are likely to be involved in several initiatives at a given time. South-South collaboration has, therefore, become a reality of the health biotechnology sector; a well-trodden route firms take in their entrepreneurial activities. Still, South-North collaborations are even more prevalent, with just over one in every two firms being active in collaboration with at least one developed country. We noted that there was a difference between the developing countries in their level of South-South entrepreneurial health biotechnology, where the countries with the smallest populations were most actively collaborating with other developing countries. This likely reflects how the small size of home markets can create the need to collaborate for the sake of a firm's viability.

Most collaborations involve linkages between the leading developing countries in health biotechnology. This survey showed the strongest linkages exist between the leading developing countries in health biotechnology, therefore reflecting how—despite distances—working together may amplify the competitiveness of relatively advanced developing countries. In addition, the results showed significant regional collaborations between firms. South Africa for example, reported active linkages with other sub-Saharan countries, and both Brazil and Cuba had active collaborations in Latin America. Our findings there indicated a dual purpose of South-South collaboration: to amplify the global competitiveness of leading developing countries in health biotechnology, and to strengthen regional ties in health biotechnology.

The health biotechnology collaborations between developing countries involve mainly end-stage commercialisation activities rather than R&D. Commercialisation activities such as distribution and marketing were by far the most common South-South collaboration activities, and more common than any research and developmental activities. This was true for all the countries surveyed in this study. The focus on end-stage commercialisation was in line with 'access to markets' being the most common reason given for South-South collaborations. South-South collaboration in health biotechnology, therefore, reflects a need for developing countries to export their products to other developing countries. The fact that the countries with the smallest populations were most active in South-South collaborations underscores this finding. Considering that some developing countries have proven track records in producing relatively affordable health biotechnology products (Thorsteinsdóttir, 2007), South-South health biotechnology collaborations may increase the availability of relatively inexpensive health biotechnology products in developing countries' markets, and the accessibility of health biotechnologies in general.

South-South collaboration is typically initiated by the participating firms themselves. The results of the survey show that little collaboration has been initiated by governmental organisations or any party other than by the participating firms themselves. International organisations and expatriates have had a limited role in encouraging South-South collaborations. Still, research on South-North collaboration has suggested that a major challenge for health biotechnology collaboration is establishing the initial linkages with the possible collaborators (Taylor *et al.,* 2007). It is likely that this challenge is experienced by the firms of developing countries as well, and our results may indicate an opportunity for greater governmental involvement. We can see in the example of the Brazil-Cuba collaboration for meningitis AC vaccine for Africa (discussed further in Chapter 4) that international organisations can play a significant role in facilitating South-South collaboration. In order to harness South-South collaboration for providing affordable options to improve health in developing countries, a further role of international organisations or philanthropic organisations is called for.

South-South entrepreneurial collaborations have a limited contribution to innovation in health biotechnology. Few of the collaborations that were reported in the survey involved knowledge creation activities tied to innovation. For example, only 13 per cent of the reported collaborations involved R&D and only 9 per cent involved clinical trials. This may reflect that many of the firms we surveyed are not active in health biotechnology innovation. Instead they may be licensing products themselves from firms that are innovators in the field, typically from the developed countries. However, we also see that some firms from China, Cuba and India have been increasingly applying their innovative capabilities to the health biotechnology field. It will be of interest to repeat the survey in the future and see if South-South collaboration will make a richer contribution towards innovation. It was also notable that collaboration involving R&D activities had a strong commercial side to it with 'joint product on market' being the most frequently cited output for the R&D collaborations. This reflects the sizable product focus of R&D collaborations, which may translate into a stronger innovation track record once more firms are able to build up innovation capacity.

From our research, we infer that developing countries' health biotechnology firms are to a significant degree relying on South-South firm collaboration as a way to expand their market potentials. Market demand has been expanding in many developing countries and it is thus an increasingly lucrative strategy to target those markets (The World Bank, 2010). To set up collaboration with local firms is an important first step towards accessing these markets. Firms in developing countries should realise that by working together they can build on each other's strengths and develop more cost-effective products. By doing so they can expand their markets considerably in the developing world where a large proportion of the population can only afford low-priced health products. Firms in developing countries can start their cooperation by focussing on marketing and distribution, but as their collaboration deepens and trust has been built up they can start to pursue further innovative activities. Still our survey shows that even though South-South firm collaboration in health biotechnology is significant, it rarely involves innovation activities. By not engaging in joint innovation, developing countries are not yet reaping the full benefits of South-South collaboration. With an increased innovation focus, developing countries can, to a larger extent, leverage their individual strengths in this field and increase the pool of resources to address their shared problems. Still, promoting a stronger innovation focus in South-South health biotechnology collaborations cannot be built solely on the backs of firms in developing countries, and supportive activities which directly target the development of health biotechnology products and services are called for from both governments in developing countries and the international community.

References

Bower, D.J. and J.C. Sulej (2007). "The Indian challenge: The evolution of a successful new global strategy in the pharmaceutical industry", *Technology Analysis & Strategic Management* 19(5): 611-624.

Chaturvedi, K., J. Chataway and D. Wield (2007). "Policy, markets and knowledge: Strategic synergies in Indian pharmaceutical firms", *Technology Analysis & Strategic Management* 19(5): 565-588.

Chaturvedi, S. (2007). "Exploring interlinkages between national and sectoral innovation systems for rapid technological catch-up: Case of Indian biopharmaceutical industry", *Technology Analysis and Strategic Management* 19(5): 643-658.

Faulkner, W., J. Senker and L. Velho (1995). *Knowledge frontiers: Public sector research and industrial innovation in biotechnology, engineering ceramics and parallel computing.* Clarendon.

Frew, S.E., R. Rezaie, S.M. Sammut, M. Ray, A.S. Daar and P.A. Singer (2007). "India's health biotech sector at a crossroads", *Nature Biotechnology* 25(4): 403-417.

Hagedoorn, J. (2002). "Inter-firm R&D partnerships: An overview of major trends and patterns since 1960", *Research Policy* 31(4): 477-492.

Kale, D. and S. Little (2007). "From imitation to innovation: The evolution of R&D capabilities and learning processes in the Indian pharmaceutical industry", *Technology Analysis & Strategic Management* 19(5): 589-609.

Lee, C.W. (2007). "Strategic alliances influence on small and medium firm performance", *Journal of Business Research* 60(7): 731-741.

Lemle, M. (2005). "Nations team up to share R&D skills in HIV/AIDS battle", *SciDev. Net.*

Maiti, R. and M. Raghavendra (2007). "Clinical trials in India", *Pharmacological Research* 56(1): 1-10.

Morel, C.M., T. Acharya, D. Broun, A. Dangi, C. Elias, N.K. Ganguly, C.A. Gardner, R.K. Gupta, J. Haycock, A.D. Heher, P.J. Hotez, H.E. Kettler, G.T. Keusch, A.F. Krattiger, F.T. Kreutz, S. Lall, K. Lee, R. Mahoney, A. Martinez-Palomo, R.A. Mashelkar, S.A. Matlin, M. Mzimba, J. Oehler, R.G. Ridley, P. Senanayake, P. Singer and M. Yun (2005). "Health innovation networks to help developing countries address neglected diseases", *Science* 309(5733): 401-404.

OECD (2006). *South-South trade: Vital for development.* Paris: Organisation for Economic Co-Operation and Development.

Pisano, G.P. (2006). *Science business: The promise, the reality, and the future of biotechnology*. Boston: Harvard Business School Press.

Ray, M., A.S. Daar, P.A. Singer and H. Thorsteinsdóttir (2009). "Globetrotting firms: Canada's health biotechnology collaborations with developing countries", *Nature Biotechnology* 27(9): 806-814.

Roijakkers, N. and J. Hagedoorn (2006). "Inter-firm R&D partnering in pharmaceutical biotechnology since 1975: Trends, patterns, and networks", *Research Policy* 35(3): 431-446.

Simonetti, R. and E. Archambault (2007). "The Dynamics of pharmaceutical patenting in India: Evidence from USPTO data", *Technology Analysis & Strategic Management* 19(5): 625-642.

Taylor, A.D., D. Brook, D. Watters, E. Dowdeswell, A.S. Daar and P.A. Singer (2007). "North-South partnerships: A study of Canadian firms", *Nature Biotechnology* 25(9): 978-979.

The World Bank (2010). *Global economic prospects: Crisis, finance, and growth*. Washington, DC: The International Bank for Reconstruction and Development and The World Bank.

Thorsteinsdóttir, H. (2007). "The role of the health system in health biotechnology in developing countries", *Technology Analysis & Strategic Management* 19(5): 659-675.

Thorsteinsdóttir, H., U. Quach, A.S. Daar and P.A. Singer (2004). "Conclusions: Promoting biotechnology innovation in developing countries", *Nature Biotechnology* 22: DC48-52. Supplement.

van Beuzekom, B. and A. Arundel (2006). *OECD biotechnology statistics—2006*. Paris: Organisation for Economic Co-Operation and Development. Available at *http://www.oecd.org/dataoecd/51/59/36760212.pdf*

Yusuf, S., K. Nabeshima and D. Perkins (2007). "China and India reshape global industrial geography", in L.A. Winters and S. Yusuf (eds.), *Dancing with giants: China, India, and the global economy*. Singapore: The World Bank and the Institute of Policy Studies.

4 Setting a Southern Course: Brazil's South-South Collaboration in Health Biotechnology

AUTHORS: *Tirso W. Sáenz, Maria Carlota de Souza-Paula, Monali Ray, Halla Thorsteinsdóttir*

4.1 Introduction

Brazil's activities in the global arena reflect a considerable emphasis on South-South collaboration. It is the largest economy in Latin America and the most populous country on the continent. Although Brazil has been independent for a relatively long time, or since 1822, it shares its history of being a colony of Portugal with several countries in Africa. Consequently, its official language being Portuguese contributes further to Brazil's affinity with Portuguese-speaking African countries. In Latin America, Brazil spearheaded the establishment of the Mercosur (or the Southern Common Market) in 1991, a regional trade arrangement initially between Brazil, Argentina, Paraguay and Uruguay. Brazil is also an emerging economy with a relatively strong economic standing among developing countries (OECD, 2009). During the latter part of the 1900s, its economy had more than its shares of ups and downs, and for some time, Brazil suffered from high inflation. It has, however, undergone strong economic growth in recent years, and is now the 10th largest economy in the world (The World Bank, 2007).

Brazil has prioritised scientific development, and in recent years, both public and private resources for scientific and technological activities have increased considerably. In 2003 it allocated 1.26 per cent of its gross domestic product (GDP) to science and technology, but in 2008 this increased to 1.43 per cent (Ministry of Science and Technology, 2010; Petherick, 2010). Brazil has a relatively long history of promoting biotechnology development. In the 1970s, the federal funding agency, Conselho Nacional de Desenvolvimento Científico e Tecnológico (CNPq, 2007), launched two programmes to support biotechnology development—the Integrated Programme on Genetics (PID) and the Integrated Programme on Tropical Diseases (PIDE). In 1981, the government set up the National Biotechnology Programme (PRONAB) to integrate activities in the different types of biotechnology. Since then further programmes have been set up to promote biotechnology development and have been supported by governments of different ideological orientations. As a result of this emphasis, Brazil has capacity in health biotechnology, including well-equipped universities and research institutes (Ferrer *et al.,* 2004). It is reflective of Brazil's strengths in the health biotechnology field that it is third ranked among all low- and middle-income countries in terms of number of papers published in international peer-reviewed journals in health biotechnology, behind China and India, but slightly ahead of Russia. Compared to other developing countries strong in health biotechnology, Brazil and South Africa stand out in publishing their research in relatively high-impact journals (Thorsteinsdóttir *et al.,* 2006). In addition, Brazil has begun to build up its indigeneous private sector in health biotechnology, pursuing process and product innovation (Rezaie *et al.,* 2008). Further, Brazil's fledgling donor efforts for South-South development are focussed on establishing technical cooperation programmes, in which research for development constitutes a growing component (Costa Vaz and Inoue, 2007). With both a bent towards South-South collaboration, and strengths in health biotechnology, there is considerable scope for Brazil to engage in South-South collaboration in health biotechnology.

Still, knowledge is missing on the extent to which Brazil has engaged in South-South collaboration in health biotechnology, and with which countries; what key challenges this collaboration has faced; and what impacts it has had so far. Further, there is more to be known about which strategies successfully cultivate South-South collaboration in the health biotechnology field. The aim of this chapter is to address these questions by presenting case-study research on Brazil's health biotechnology collaboration with Argentina and Cuba (Box 4.1). We start by mapping the main collaborations Brazil has had in the health biotechnology field, focussing both on research and entrepreneurial collaborations. We then discuss in more detail some of the experiences Brazil has had in collaborating with Argentina and Cuba, and present conclusions based on our case-study research.

BOX 4.1

Case-study research on Brazil's collaborations: Methodology

To understand Brazil's South-South collaboration and its regional role in the health biotechnology field we carried out case-study research on its partnerships with Argentina and Cuba. We studied a total of two cases and defined each case as the bilateral collaboration between Brazil and Argentina *versus* Brazil and Cuba. We chose to focus on collaboration with Argentina because it is Brazil's main Southern collaborator in the health biotechnology field and with Cuba because it has had considerable entrepreneurial collaboration with Brazil.

Both Argentina and Cuba have placed emphasis on the development of health biotechnology. In the early 1980s, Argentina began establishing biotechnology research institutes, and now has a number of universities, public research organisations, and firms active in the field (Carullo 2005; Niosi and Reid, 2007). Its publication record is relatively strong in the field, but its patenting record is weaker as almost no biotechnology patents have been granted to Argentina (Niosi and Reid, 2007). Cuba also started to invest in its biotechnology sector in the early 1980s and built up both human resources and innovation infrastructure. Its publication record in the field is limited but it has a number of biotechnology patents (Quach *et al.,* 2006). Cuba is also one of the few developing countries able to develop new-to-the-world innovation in the science-intensive biotechnology sector and harness these innovations for improving the health of its population.

As described in Chapter 1, the case studies relied on multiple sources of data, including scientometric analysis of co-publications, a survey of firms about their collaborations, document analysis of policies and background literature, and 42 interviews with experts in Brazil, Argentina and Cuba (Table 4.1).

TABLE 4.1

Breakdown of number of interviewees in case-study countries

Countries	Number of interviewees
Brazil	15
Argentina	17
Cuba	10
Total	42

4.2 Government interest and support

In the last two decades, the government of Brazil has promoted South-South collaboration as an integral part of its foreign policy. While the main focus of these efforts has been on commercial relations, it has moved beyond that, and is increasingly involving S&T (Lafer, 2000; Mathias, 2002; Saraiva, 2007; Soares de Lima, 2005). Brazil spearheaded initiatives for regional integration that resulted in the creation of Mercosur (or the Southern Common Market) in 1991, which aimed to facilitate trade and the movement of people and currency between its member states of Argentina, Brazil, Paraguay and Uruguay. Other countries have since joined as either associate or full members. Mercosur fosters S&T cooperation especially through RECYT (the Mercosur Permanent Committee for S&T) and has recently developed a framework to harness joint capacity in S&T to address fundamental problems, and make 'science and technology, in the long term, a conscious, vital and genuine component of Latin American culture.' (Marcano González, 2006). In conjunction with the European Union, Mercosur has also established Biotecsur, a platform designed to focus on priority themes in biotechnology for the region.

Brazil is also active in multilateral programmes outside of Latin America, and promoted the establishment of the trilateral IBSA network in 2004 (IBSA, 2005 and 2007; Mokoena, 2007; Puri, 2007). IBSA encourages both trade relations and S&T collaborations in key sectors such as energy and health (see also Chapters 7, 8 and 9 for discussion of IBSA). It is a developmental initiative for promoting South-South cooperation and exchange. It encourages linkages between the participating countries in several areas including energy, education, health, trade, and S&T. The participating countries have decided on priority health areas to focus on and have divided the leaderships of the areas between themselves. Brazil is leading research cooperation in malaria as a priority health area, South Africa is leading tuberculosis initiatives, whereas India is taking the lead in research cooperation on HIV/AIDS. Other modes of cooperation in S&T will include specialised scientific exchanges, training and workshops in areas of national strength, information exchange on traditional knowledge, and collaboration in basic R&D. In addition to collaborating on research activities, the countries also share information on arrangements for technology transfer and intellectual property rights' issues and on arrangements of protecting biodiversity and traditional knowledge. The IBSA agreement thus signals Brazil's focus on increasing its relations with emerging economies, both to strengthen trade and harness science to address shared problems.

In promoting South-South collaboration in S&T, Brazil has also set up its own programmes and agreements. These include PROSUL, which promotes partnerships in S&T within South America, and PROAFRICA, which focusses on promoting S&T capacity building in the Portuguese-speaking African countries of Angola, Cape Verde, Guinea-Bissau, and Mozambique (CNPq, 2007). The funding to these programmes is at low levels with, for example, US $4.5 million being allocated to PROSUL from 2003 to 2006. Although these programmes have low levels of funding, they represent the Brazilian will to promote South-South cooperation, and also to work with countries that have limited S&T capacity. Additionally, the Brazilian government has a number of bilateral agreements to encourage S&T collaboration with developing countries. The federal funding agencies, CNPq and Coordenação de Aperfeiçoamento de Pessoal de Nível Superior (CAPES, Brasilia), have played a significant role in the signing of agreements and encouraging this collaboration. They are mostly in Latin America and include cooperation agreements with Argentina and Cuba. In addition CNPq has an agreement with TWAS, the Academy of Sciences for the Developing World, and offers post-doctoral fellow positions for developing countries' researchers to receive training in Brazil.

When it comes to governmental promotion of South-South collaboration in biotechnology, including health biotechnology, Centro Brasileiro-Argentino de Biotecnologia (CBAB/CABBIO) is Brazil's main programme (Assad *et al.,* 2000; Dellacha, 2003; Guimarães, 2000; Souza-Paula *et al.,*

2005). It was established in 1986 and its goal was to create a virtual centre to foster training and joint research between Brazil and Argentina. It was a part of an initiative to encourage collaboration between the two countries in 12 fields, including nuclear energy, space research, and communications. CBAB/CABBIO is strongly focussed on developing human resources—it has organised around 300 courses for thousands of students—and by 2008 it had supported 100 research projects. It prioritises basic research fields, like microbiology and genetics, as well as certain applied health areas, such as hepatitis B vaccines or Chagas diagnostics. CBAB/CABBIO requires equal funding from both partners and together they allocated around US $15 million for the period 1987 to 2002 to research collaboration and capacity-building activities (Souza-Paula, 2005).

Brazil has more recently emphasised Cuba as a collaborator, particularly in the health biotechnology sector. The two countries re-established diplomatic relations in 1986, and their first biotechnology agreement was signed in 1993. In the last decade, their governments signed several agreements to work together, both on research collaboration and on developing and producing health products. They have designated their foremost public research organisations—Centro de Ingeniería Genética y Biotecnología (CIGB, Havana) and Oswaldo Cruz Foundation (Fiocruz, Rio de Janeiro)—to work on developing products such as human erythropoietin (EPO), Dengue fever, and α and β pegylated interferon, among others.

The Brazilian government, thus, has made it clear that it takes South-South collaboration seriously and puts collaboration with developing countries firmly on its economic and S&T agendas. Its initial focus was on other countries in Latin America, but it is now expanding its horizon to emerging economies, such as India and South Africa, and Portuguese-speaking countries in Africa.

4.3 The geography of Brazil's health biotechnology collaboration

4.3.1 Mapping research collaboration

To examine the levels and patterns of Brazil's collaboration with other low- and middle-income countries we mapped its co-authorships with those countries. We identified papers from the Scopus database for the period between 1996 and 2009 (see Chapter 1 for details of the methodology). We identified 1,021 papers that Brazil has co-authored with other developing countries for the period studied. As discussed in Chapter 2, compared to other low- and middle-income countries, Brazil has the most South-South co-authored papers. The second-ranked country in terms of number of South-South co-authored papers is China, with around 970 papers. Seven of Brazil's top 10 partners are in the Americas, with the three countries outside this region being China, India and South Africa (Figure 4.1). The heavy emphasis on collaboration with other Latin American countries reflects Brazil's regional emphasis on South-South collaboration. The observation that the three countries outside the region are the so-called emerging economies reflects that Brazil has indeed a growing focus on collaboration with those countries.

FIGURE 4.1

The main developing countries Brazil collaborates with in South-South health biotechnology research, based on the number of co-authored papers between 1996–2009

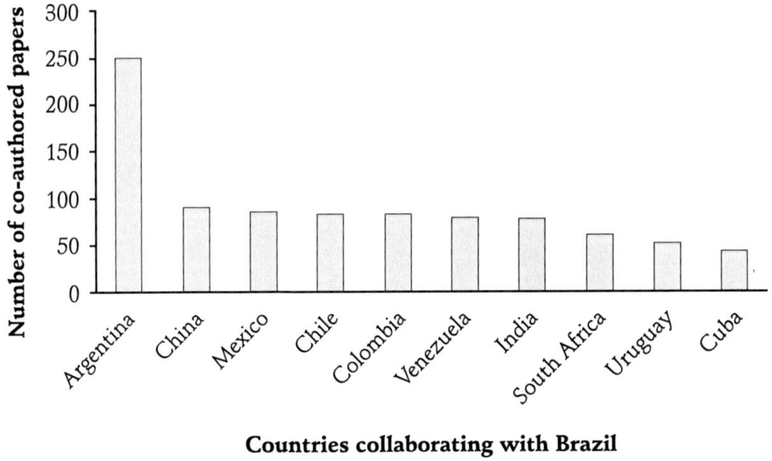

Source: Authors' presentation of data compiled by Science-Metrix using Scopus.

Among developing countries, it is also obvious that Argentina is Brazil's main research collaborator in health biotechnology (Figure 4.1), and that these two countries co-author more South-South papers in this field than any other countries (see Chapter 2). Brazilian authors published 250 papers with Argentinean authors during the period, *versus* 91 papers with

Chinese and 86 papers with Mexican researchers, who were the second and third most frequent publishing partners for Brazil. In comparison Brazil and Cuba co-authored 43 papers during the period.

When we looked at the subfields of Brazil's collaborations with developing countries in general, we saw that 'Genetics and Heredity' is by far the most common subfield of these collaborations. This finding is consistent with results on South-South collaboration in general which show that 'Genetics and Heredity' is the most common subfield of South-South collaboration (see Chapter 2). The traditional biotechnology subfields, 'Biochemistry and Molecular Biology,' 'Biomedical Engineering,' and 'Microbiology' then ranked in second to fourth place. 'Parasitology' was in sixth place and 'Tropical Medicine' in eighth place reflecting a relatively moderate emphasis on collaboration that is likely to be focussed on health problems that developing countries predominantly suffer from.

We also looked at the extent to which Brazil collaborates with high-income countries in the health biotechnology field. For the whole period studied, 1996 to 2009, Brazil collaborated around five times more with Northern countries than with Southern countries with 1,021 papers in South-South collaboration *versus* 5,121 co-authored papers in South-North collaboration. It is not surprising that Brazil has more collaboration with Northern countries than Southern countries as most of the knowledge production in the health biotechnology field is in high-income countries (see Chapter 2). Still around 18 per cent of Brazil's international collaboration in health biotechnology is with other developing countries for the whole period studied, increasing from 16 per cent at the beginning of the period to 19 per cent at the end. Brazil has significantly fewer co-publications with high-income countries than China has, with almost 18,000 such co-authored publications, but more than India which has just over 4,000 such papers co-authored for the period studied. Brazil's main Northern collaborators are the United States (2,705 co-authored papers), United Kingdom (746 papers), France (677 papers), and Germany (502 papers). Research on Brazil's international scientific collaboration in general supports our findings and shows that Brazil has significantly greater collaboration with high-income countries than other low- and middle-income countries, with the United States and European countries being their main partners (Narváez-Berthelemot *et al.,* 1999).

4.3.2 Mapping entrepreneurial collaboration

To examine Brazil's entrepreneurial collaboration with other developing countries we administered a survey of health biotechnology firms in Brazil and other leading developing countries in the health biotechnology field (see Chapter 1 for details of the methodology). Looking at the countries that have at least three entrepreneurial collaborations with Brazil, we can see that, again, Brazilian firms have the most ties with other countries in the Americas (Figure 4.2). The only other countries on this list are China and India, again reflecting growing ties with emerging economies. Argentina seems to be Brazil's primary entrepreneurial collaborator, but with collaboration levels comparable to those of China and Cuba. While most of Brazil's entrepreneurial collaborations only involve marketing and distribution activities, its main R&D collaboration appears to be with Cuba. As R&D activities relate closely to the innovation potential of heath biotechnology projects, we were keen to study these further.

In our study, therefore, we choose to focus on Brazil-Argentina collaborations because of their high level of joint research. We then

FIGURE 4.2

The main developing countries Brazil collaborates with in South-South entrepreneurial collaborations in health biotechnology, based on our survey of firms

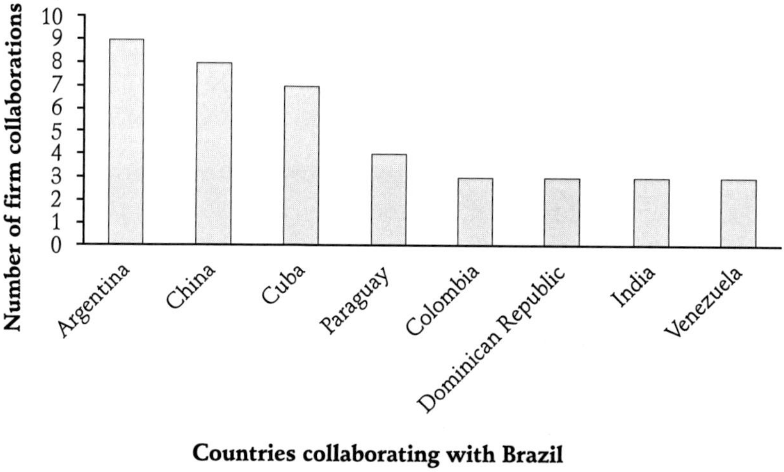

Source: Authors' presentation of their own survey data.

focussed on collaborations between Brazil and Cuba because Cuba is the most common partner for Brazil's entrepreneurial R&D collaboration with the potential to lead to innovation. The governments of both Argentina and Cuba have placed an emphasis on developing their biotechnology sectors (Asociación Redes-Biotech, 2005). In the early 1980s, Argentina began establishing biotechnology research institutes, and now it has approximately 60 centres and 35 firms in the field (Carullo, 2005; Niosi and Reid, 2007). It holds biotechnology as a strategic axis of its present S&T policy. During the economic crisis in Argentina in the mid-1990s, the emphasis on promoting biotechnology diminished, but was kept alive by CBAB/CABBIO and via collaborative linkages with high-income countries. According to interview evidence, Argentina seems to lack a focus on South-South collaboration in biotechnology, with the exception of CBAB/CABBIO. It has signed more than 150 inter-governmental and inter-institutional agreements to promote international S&T collaborations in general. They include collaboration with a number of developing countries, particularly in Latin America but no particular emphasis is placed by Argentina on collaboration with Southern countries. Still over 24 per cent of its internationally co-authored papers in the health biotechnology field is with other low/middle-income countries. Brazil is by far Argentina's most common Southern partner with 250 co-authored papers from 1996 to 2009, but another neighbouring country, Chile, comes in second place with 73 such papers. Still its main collaborator was the United States with whom Argentina co-authored 864 health biotechnology papers, next was Spain (350 co-authored papers), and France (268 co-authored papers). Brazil ranked fourth globally as collaborator with Argentina.

Cuba has prioritised biotechnology development since the early 1980s, with its quest to develop interferon, and continued to do so during the economic recession of the early 1990s. It has built up significant institutional infrastructure in the field with research, developmental, manufacturing, and entrepreneurial capacities (Sáenz, 2008; Thorsteinsdóttir *et al.,* 2004). It has been one of the most successful developing countries at creating new-to-the world, innovative biotechnology products, exemplified by its own meningitis B vaccine, synthetic Haemophilus influenzae type vaccine, and therapeutic cancer vaccine, and exports health biotechnology products to a number of countries (Sáenz, 2005; Thorsteinsdóttir, 2007). An integral part of Cuban health biotechnology is a focus on local health needs and close ties between the public health sector and the science system and it has created the Western Havana Scientific Pole (WHSP), a cluster of scientific research and technology institutions in the capital to promote strong ties between research institutes, hospitals, government agencies, and the University of Havana (Lage, 2007 and 2008; Thorsteinsdóttir *et al.,* 2004b). Research from the Cuban organisation Biomundi (Havana) specifically examined co-authorship between Brazilian authors and Cuban WHSP authors and showed a steep increase in co-publications in the last few years or since 2007. Their analysis, however, was not confined to the health biotechnology field.

Cuba has placed political and economic emphasis on South-South collaboration, and has established formal S&T collaborations with a large number of developing countries in Latin America, Asia and Africa. South-South collaboration is, according to interviews with experts, closely aligned with Cuba's foreign policy and its emphasis on trading with Southern countries. In 2008 it had scientific and technological agreements with 32 developing countries *versus* 11 agreements with European countries, Canada and Australia, reflecting a stronger emphasis on collaboration with Southern countries. It reflects this emphasis that over 33 per cent of its internationally co-authored papers from 1996 to 2009 in health biotechnology are with other low/middle-income countries. Spain is Cuba's most common collaborator (130 papers), Mexico ranks second (54 papers), Germany-third (48 papers), Brazil-fourth (43 papers), and the United States-fifth (42 papers). Cuba is the only developing country that is active in the health biotechnology sector that does not have the United States as its foremost collaborator. This is without a doubt due to the effects of the US trade embargo against Cuba which limits scientific collaborations between the two countries.

4.4 Collaboration in research activities

Brazil and Argentina have a long-standing tradition of research collaboration. It started in the early 1900s in biology and has been strengthened more recently by the Mercosur, CBAB/CABBIO, and PROSUL programmes, amongst others (Velho, 1997). Some examples of Brazil-Argentina collaboration are the collaboration between the University of São Paulo, Brazil, with National University of the Littoral, Argentina in diagnosis of Trypanosoma cruzi, the pathogen that causes Chagas disease; collaboration between the University of São Paulo with the University of Buenos Aires, Argentina in characterisation of the developmentally regulated amino acid transporter gene from Leishmania amazonensis; collaboration of the University of Belo Horizonte, Brazil and Fiocruz, Brazil with the National University of Cordoba, Argentina on the characterisation of the Schistosoma mansoni parasite; and the Paraná Molecular Biology Institute with National University of Misiones, Argentina on hantavirus.

Brazil's research collaboration with Cuba does not have as strong a foundation; still, the two countries have set up several government initiatives to promote and plan collaborations since the late 1990s. Some examples of Brazil-Cuba collaborations are between the University of São Paulo, Brazil and the Cuban Center of Marine Bioproducts on characterisation of compounds with potential medical significance; the University of São Paulo and the Center of Genetic Engineering and Biotechnology, Cuba and the University of Oriente, Cuba on the genetics of Xanthomonadales bacteria; and the Federal University of Rio de Janeiro, Brazil and the Center of Molecular Immunology, Cuba on techniques for high yield purification of EPO.

4.4.1 Reasons for the collaboration

Our analysis revealed several factors motivating collaboration:

1. *Complementary scientific strengths:* All three countries are strong in health biotechnology, compared to others in the region. As mentioned above, Brazil's reputation in this field is reflected in its track record of publications in international peer-reviewed journals, and by its considerable research infrastructure investments over the last two decades. Brazilian, Argentinean, and Cuban interviewees all emphasised complementary strengths as a driver of their collaboration. By working together, researchers were able to access each other's expertise and share knowledge. As described by a former Brazilian Minister of Science and Technology, 'Brazil

and Argentina are at the same [health biotechnology] development level and [their] integration promises the sum of potential and forces.' (Brasil Ministério de Saúde, 2005a and 2005b). Likewise, Argentina and Cuba have both built up their human resources in health biotechnology and, together with Brazil and Mexico, have the highest numbers of researchers in Latin America and the Caribbean countries.

External conditions have also stimulated the exchange of scientific strengths. Our interviewees stressed that international collaboration with Brazil became even more important for researchers in Argentina during the economic hardship of the late 1990s and the early 2000s, because it provided much-needed access to funding and research infrastructure. For some Cuban researchers, collaboration was important for gaining access to research infrastructure unattainable in Cuba as a result of the US trade embargo in effect since 1961. Our research therefore shows that the complementary strengths that the collaborators contribute to the collaboration can be of various types, from specific expertise, particular research samples, to research infrastructure.

2. *Geographical proximity:* Brazil and Argentina are the two largest countries in South America in terms of land mass. They share large common borders, which creates both needs and opportunities for their joint collaborations. Shared ecosystems and natural resources, the flow of people and diseases, and opportunities for affordable and convenient travel by investigators all encourage collaboration. Their proximity further stimulates political emphasis on partnerships in various fields, from trade to science. Programmes like CBAB/CABBIO have led to closer collaboration in health biotechnology research than with other developing countries, and are likely to explain why Brazil and Argentina publish more joint papers in this field than any other developing countries. Similarly, the fact that there is considerable distance between Cuba and Brazil has deterred their collaboration as it has at times been challenging, particularly for Cuban researchers, to get the financial resources to allow them to visit each other.

3. *Shared health problems:* Brazilian and Argentinean researchers further emphasised that shared health problems were an important reason for their research collaboration. When we looked at the subjects of their co-authored papers in health biotechnology we clearly could see that this collaboration has a strong focus on common problems such as Chagas disease (Box 4.2). By working together they can combine their expertise, samples, and other

BOX 4.2

Latin America united against regional health hazard

Chagas disease, caused by the protozoan Trypanosoma cruzi, is a major public health problem in Latin America. Transmitted by hematophagous bugs that colonise the homes of poorer rural communities in Latin American countries, it is a potentially life threatening illness that causes the progressive destruction of heart muscle. In the 1980s, an initiative was mounted by a network of scientific laboratories in 19 Latin American countries to conduct research towards a reliable, rapid ELISA diagnostic test for Chagas disease; one that would be tailored to low-resource settings. The project received support from Spain's CYTED programme, funding agencies such as CNPq and FAPESP in Brazil, and FONCyT and UBACyT in Argentina, the bilateral CBAB/CABBIO programme, as well as the World Health Organization (WHO) and is thus an example of South-South-North collaboration (Bialy, 1998; Ponce *et al.,* 2005; Umezawa *et al.,* 2003).

In 1990, the network of labs selected T. cruzi antigens cloned by researchers in Argentina, Brazil, Spain and Venezuela to use for the ELISA test. To aid the selection, the group relied on software developed at Fiocruz, and pilot-scale production of the recombinant antigens was performed in Brazil. In 2005, scientists from Brazil, Argentina, Venezuela, El Salvador, Nicaragua and Honduras, working with US and Spanish partners, evaluated the performance of their T. cruzi multi-antigen ELISA diagnostic test in Honduran, El Salvadorian, and Nicaraguan laboratories. The results showed 99.6 per cent sensitivity and 99.9 per cent specificity (Ponce *et al.,* 2005). Trademarked Chagas Stat-Pak Assay, the T. cruzi ELISA diagnostic test, is now being commercialised by the US biotechnology firm, ChemBio Diagnostic Systems Inc. (New York). In 2006, the firm fulfilled a contract from the WHO for a Chagas disease screening programme for children in endemic regions of Bolivia.

resources needed to carry out research on their shared health problems and thereby strengthen their capacity to do the needed research. The two countries set up a joint committee to identify urgent health problems which prioritised work on anti-retrovirals, Chagas disease, and tuberculosis, primarily, but also antigens against Leishmania amazonensis, the genome of Leptospirosis, and Hantavirus. Shared health problems were also a driving force for Brazil-Cuba research collaborations, but they have also focussed their attention on cooperation more generally in nanotechnology, informatics and energy.

4. *Scientific diaspora:* Many well-trained scientists in developing countries are part of the so-called scientific diaspora, leaving their countries for opportunities—usually in the north—or for political reasons. Several Brazil-Argentina collaborations involve the Argentinean diaspora in Brazil where the collaboration allowed the Argentinean researchers who had moved to Brazil to continue to work with partners in Argentina. Although the diaspora can be seen as eroding for the leaving country's S&T, here, it has strengthened bi-national collaboration between Brazil and Argentina. In contrast, Brazil-Cuba collaborations did not seem to involve a scientific diaspora in either country and the scientific diaspora were thus not a driving force for their collaboration.

4.4.2 Challenges of the collaborations

The main challenge to research collaborations discussed by our interviewees was a lack of funding for research in general and thus also for their research collaboration. Although both Argentina and Cuba are strong in biotechnology research, limited finances often affect their ability to collaborate. There were times when CBAB/CABBIO was the main funding source for Argentinean biotechnology researchers, which led to pressure to diversify the projects and pulverise funding. Some Argentinean researchers decided not to collaborate with Brazil—even when their interest was great—in favour of pursuing projects with developed countries, where funding and access to infrastructure and services were better. In recent years, new initiatives have increased funding for S&T in Argentina, but researchers still consider the limitations severe. For Cubans, funding restrictions were problematic and some of the agreements made to strengthen S&T collaboration were not fully realised. Instead of joint funding, some of the projects are reportedly funded entirely by the Brazilian side, which weakens collaborations, particularly collaboration involving universities in Cuba. Even so, recent unpublished research by Biomundi shows a steep increase in Cuban-Brazil co-publications in the last few years, totalling more than 20 publications in 2007. Conversely, Brazil's funding for research has grown extensively in recent decades, and more money is being directed towards South-South collaboration than ever before. This creates favourable conditions for Brazil to promote cooperation.

According to our interviewees, other challenges in health biotechnology collaborations have been the bureaucracy involved in obtaining what limited and uncertain support exists, delays in payments, and restrictions of the number of funded projects by each institution. In some cases, researchers decided not to seek support through programmes like CBAB/CABBIO or PROSUL because the process was perceived to be burdensome. Instead, they relied on their own research funding for their collaborations. This could lead to unequal collaborations when one partner has more resources for the research than the other. It also can make it more challenging to coordinate research activities when partners are at different places in their funding cycles. Access to joint funding programmes is therefore key to fostering research collaboration and to cultivating longer term collaborative projects.

4.4.3 Impacts of the collaborations

We also asked researchers to specify the main impacts of their collaborations. Closer knowledge flow between the countries and increased publications in international peer-reviewed journals were clear outcomes identified by many of our interviewees. The fact that Brazil and Argentina publish the highest number of jointly written health biotechnology papers of any South-South partners likely reflects the success of establishing specific programmes and funding to encourage these collaborations, which governments in developing countries should pay attention to. The CBAB/CABBIO programme and other specific resources for regional collaboration are likely to have encouraged close collaboration between Brazil and

Argentina. Also the limited resources allocated to Brazil-Cuba collaboration have also been key to their research collaboration. One of the impacts of this collaboration is also the potential to remain scientifically active during economic hardships. This outcome was particularly emphasised by Argentinean researchers as a result of the collaboration with Brazil during the recession in Argentina in early 2000s. It is noteworthy that, despite the reduction in resources allocated to research during Argentina's economic difficulties, the country continued to publish papers in international peer-reviewed health biotechnology journals. In 2008, for example, it produced 452 papers, close to double its 1996 level of 221.

4.5 Collaboration in entrepreneurial activities

In contrast to the active research collaboration between Brazil and Argentina, entrepreneurial collaboration has not been as active, and has not applied new knowledge to the development of product and services. This is partly because entrepreneurial collaborations between Brazil and Argentina are limited in scope and focussed almost entirely on marketing activities. Although health systems representatives in Brazil and Argentina are discussing potential collaborations in developing health products, they are still at the negotiation stage. There have been some attempts by firms in the two countries to collaborate on developmental activities, but so far the outcomes have been minimal. Other research confirms our findings; Alcorta *et al.* (1997), for example, have found low collaboration rates and limited contributions to innovation in their study of Mercosur firms.

Our research revealed more examples of innovative entrepreneurial collaborations between Brazil and Cuba than with Argentina. This is partly because Cuban health biotechnology has a particularly strong innovation record (Thorsteinsdóttir *et al.,* 2007). Much of the discussion in this section, therefore, draws upon the experiences of the collaborations between Brazil and Cuba. Interestingly, almost all cases of Brazil-Cuba collaboration involve the entrepreneurial arms of public research organisations, and are driven by specific governmental agreements. Bio-Manguinhos (Rio de Janeiro, Brazil) the commercial arm of Fiocruz, for example, collaborates with Heber Biotec (Havana, Cuba) the commercial arm of CIGB (Havana). This is to be expected in the case of Cuba where entrepreneurial health biotechnology is entirely carried out by governmental organisations. An exception to this pattern is the collaboration between the public Cuban research organisation CIM (Havana) with the private Brazilian firm Eurofarma Laboratórios (São Paulo). It involves running clinical trials on nimotuzumab, a new Cuban biotechnology medication for several types of cancer (see Box 3.1 in Chapter 3).

4.5.1 Reasons for the collaborations

The research pointed to a number of reasons for the entrepreneurial collaborations:

1. *Technology transfer:* As mentioned above, Cuba has a relatively strong innovation record, but has also been active in producing affordable biosimilars to supply its health systems and gain export earnings. Brazil, too, wants to become less dependent on health biotechnology imports, and has in recent years signed technology transfer agreements with Cuban organisations to improve its local technological and manufacturing abilities. For example, Bio-Manguinhos signed a technology transfer agreements with Heber Biotec in 2003 for the production of interferon alpha-2b, and with CIMAB (Havana) in 2004 for the production of EPO. Both agreements stipulate that the technology transfer to Brazil will be carried out in stages starting with the Cuban institutions supplying packaged finished Cuban product to Bio-Manguinhos to be sold under the latter's trade mark, next the packaging will be at Bio-Manguinhos and by the end, Bio-Manguinhos will wholly produce the biologics and pay royalties to the Cuban organisations. The driving force for Brazil in these agreements is the wish to reduce its dependency on imports so as to lower prices and accelerate the public's access to needed products. The main benefit to Brazil is thus savings to the public health system but to Cuba it is income earnings from royalties.

2. *Complementary expertise:* Brazil and Cuba have also collaborated because they have been able to contribute their respective strengths to develop and manufacture health products. One of the best

examples of South-South collaborations fuelled by complementary expertise is the collaboration on producing a bivalent AC meningitis vaccine for African countries, following a request from the WHO (Box 4.3). This collaboration draws on the respective strengths of both Cuba and Brazil: the Finlay Institute (Havana) synthesises the active ingredients for the vaccine, and Bio-Manguinhos produces it,

BOX 4.3

Vaccines for Africa's meningitis belt

To counter a meningitis outbreak in 2007 in the so-called 'meningitis belt' of Africa, the WHO decided to assess the status and production capacity of polysaccharide vaccine manufacturers worldwide. This examination identified Bio-Manguinhos, in collaboration with the Finlay Institute, as the most suitable supplier. They could quickly provide the needed products at a lower price than alternative suppliers. The meningitis belt in Africa stretches from Senegal in the west to Ethiopia in the east and covers several low-income countries. Its population is estimated to be around 300 million. Samples from meningitis-infected individuals showed that the cases were caused by Neisseria meningitidis serogroup A, which is the most common serogroup in Africa, but exists in neither Brazil nor Cuba.

The Finlay Institute has had a long history of meningitis research and managed to control a meningitis outbreak in Cuba in the mid-1980s by developing a first-of-its-kind vaccine worldwide. Bio-Manguinhos also has extensive experience in vaccine research and manufacturing, and has developed an efficient scale-up process using lyophilisation. By relying on their respective strengths, these two organisations were able to react in a timely fashion to the meningitis outbreak in Africa and supply WHO with the needed vaccine. According to the Finlay Institute, between 2007 and 2009, some 19 million doses were produced and distributed in Burkina Faso, Ethiopia, Mali and Nigeria. The vaccine's price is much lower than on the international market, and lower than would be possible without cooperation. Neither organisation alone would have been able to respond so quickly and efficiently to this request. This, therefore, demonstrates how South-South collaboration can be harnessed to address a health threat spurred by demand and funding from an international organisation (WHO 2007a and 2007b).

using its strengths in the lyophilisation process. Another example of complementary expertise is the joint development of pegylated interferon by Bio-Manguinhos and CIGB. Pegylated interferon involves slow release of the drug delivery process, thereby reducing the number of injections required of the interferon. The collaboration relies on extensive expertise in interferon by CIGB and the expertise in the lyophilisation process offered by Bio-Manguinhos necessary for vaccine scale-up and manufacture.

3. *Access to international markets:* Access to each other's markets was another important motivator for collaborations. Brazil has the largest pharmaceutical market of these three countries because of its large population size, increasing middle-class population, and improved access to health care. Market access was also the main reason for Brazil's limited entrepreneurial collaborations with Argentina. It was a particularly important reason for Cuba's entrepreneurial collaboration with Brazil and other developing countries, in light of the US trade embargo. As discussed in Chapter 3, the survey results confirmed the importance of access to markets as motivation for the collaboration and the entrepreneurial organisations in Brazil, and Cuba gave it as the main reason for their collaborations. What is interesting, as we discussed in Chapter 3, is that the R&D collaborations that we observed between the countries we focussed on often also include marketing collaborations. These different types of collaboration activities are therefore not mutually exclusive.

4.5.2 Challenges of the collaborations

The main limitation of the entrepreneurial collaborations appears to be a weakness in joint development activities, especially by Brazilian and Argentinean firms. While their research collaboration has led to knowledge that can be further developed and commercialised, as was the case in the collaboration on Chagas diagnostics (Box 4.2), overall, entrepreneurial collaboration is still relatively limited and does not align well with their research collaboration. The commercialisation of the Chagas diagnostic kit, for example, is not being carried out by Brazilian or Argentinean firms, but instead by a US firm. The reason for the limited success of entrepreneurial collaboration between Brazil and Argentina could be partly due to weaknesses in their health biotechnology private sectors: firms are young, they do not engage much in product development, they have relatively weak patenting records, and they are not prepared to take the risk of co-development with firms in other countries.

Another reason for this lack of entrepreneurial collaboration could be the limited extent to which the programmes aimed at supporting S&T collaboration have focussed on firm involvement. For example, CBAB/CABBIO was initially created with an emphasis on fostering technological and entrepreneurial collaborations; however, in practice, it had a much stronger focus on capacity building. Some early projects under CBAB/CABBIO involved firm collaborations, but did not succeed for other reasons, including a lack of financing, intellectual property issues, and barriers to exchanging research material. Other approaches to promote cooperation were mainly centred on human resources and research.

The more active Brazil-Cuba entrepreneurial collaborations have experienced some difficulties, including a case in which a difference in opinion on intellectual property rights led to the termination of the collaboration. Challenges also arise related to other legal aspects, including certification, control, and approval of health product norms and clinical trials across countries. To address these types of challenges, a formal cooperation between the drug regulatory agencies Agência Nacional de Vigilância Sanitária (ANVISA, Brasilia) and the State Center for Quality Control of Medicines (CECMED, Havana) was successfully established. Because of the WHO's role in the Africa meningitis project—including prequalification of the meningitis AC vaccine—the agencies were given an extra push to collaborate and had more funding available for this purpose than they would have had otherwise. At the beginning of 2008, the two governments signed a formal agreement, stipulating that ANVISA and CECMED exchange documentation and visits, and work towards coordinating the registration of their countries' medical products. As a result, the agencies created an Inter-Institutional Regulatory Commission. This cooperation has already led to increased clarity about each other's systems.

4.5.3 Impacts of the collaborations

Brazil has promoted technology transfer arrangements from Cuba to help increase its capacity to produce biosimilars. This reduces both dependence on international imports, and leads to the availability of less expensive health products. For example, the international market price of 2,000 units of EPO is US $50, but through the collaboration with Cuba, Bio-Manguinhos can now produce the same quantity for only US $3, resulting in significant savings for the Brazilian health system. For Cuba, this technology transfer agreement with Brazil has led to increased revenue by supplying the active pharmaceutical ingredient to Bio-Manguinhos. Once the technology transfer process is complete, the Cuban organisations will also receive royalties from Brazil for access to their techniques.

A further impact of this collaboration has been the increased availability of vaccines for African countries dealing with meningitis, and at a lower price compared to vaccines on the international market. Multinational pharmaceutical companies produce a conjugated meningitis vaccine with a high immunisation period that can be used for meningitis AC for children. However, this vaccine is relatively new and costly, listed for over US $80 a dose according to the US Centers for Disease Control and Prevention (CDC, 2009). It provides protection against several serogroups of meningitis which increases its price, but two of these serogroups (Y and W135) are not found in Africa. It also increases the price of the vaccine that it requires two or three doses for it to provide an immunisation. Administering multiple doses of the vaccine to reach immunity is challenging in poor countries, which typically do not have the infrastructure to be able to carry out a complicated vaccination campaign. The vaccine produced by the Brazil-Cuba collaboration is a polysaccharide which does only have a short immunisation period of around three years and does not cover as wide of a spectrum of serogroups of meningitis. It, however, costs only US $0.95 per dose and is proven to be effective to control the epidemic which is the most prevalent in Africa. In comparison a polysaccharide A, C, W135 and Y vaccine produced by multinational companies costs over US $15 a dose (Grogg, 2010).

This case demonstrates that South-South collaboration can leverage strengths in different developing countries while providing more affordable health solutions to others. As discussed above, to support this undertaking, the regulatory agencies in both Brazil and Cuba started to work together, and in 2008, their governments signed a formal agreement to do so. This cooperation led to increased clarity about each other's systems. Because of the WHO's role in the meningitis project for Africa, the organisations have been encouraged to collaborate and exchange information and access to more funding than they otherwise would have.

4.6 Conclusions

Our case-study research on Brazil's health biotechnology collaborations with Argentina and Cuba has identified a strong potential for South-South collaboration and shed light on approaches to strengthen it. The main messages from this research are:

Governments can cultivate close linkages between researchers in developing countries by setting up programmes to support their collaborations, and allocating dedicated funding to them. The CBAB/CABBIO programme between Brazil and Argentina appears to have led to the highest number of co-publications in international peer-reviewed journals from any South-South partnership. More recently, the Brazilian and Cuban governments have played an important role in cultivating joint innovation between the two countries. By setting up channels for collaborations and providing financial support, governments can play a leading role in fostering linkages between developing countries. However, it should be recognised that active research collaboration is unlikely to lead to either co-development or innovation without a concerted effort aimed at encouraging such activities, and the active involvement of firms and other entrepreneurial organisations.

There is significant scope for technology transfer arrangements between developing countries. As more developing countries build capacity in health biotechnology, they can increasingly share these capabilities with other developing countries, thereby building capacity in this field. Our research shows how public research organisations and their entrepreneurial arms have worked together to transfer technologies between Cuba and Brazil. This arrangement may also be beneficial for collaboration between private-sector firms. Increased South-South technology transfer can lead to strengthened local production, increased income from royalties, and price reductions for the health systems of developing countries.

The role of international organisations can be of paramount importance for fostering South-South collaboration in health biotechnology. The support of international organisations, such as WHO, can be important in encouraging South-South collaboration and overcoming financial obstacles to South-South collaboration; a hurdle that often restricts countries from leveraging their strengths. As we observed in the case of Brazil's and Cuba's collaboration on the meningitis vaccine, such support can also strengthen the involvement of regulatory and other organisations in developing countries by cultivating and aligning collaborative agreements which, in turn, can strengthen their innovation potentials.

Some developing countries can work together to secure timely and affordable health products for improving global health. Developing countries can leverage their strengths in this field jointly to supply health products for epidemics in the developing world. International organisations and philanthropic organisations should place a stronger emphasis on drawing upon the strengths and potential of developing countries to work together when supplying countries in need with health products. As developing countries' health products are usually less expensive than those from the north, savings would be realised by the recipient countries that need them most.

This case study shows that there are great opportunities for strengthening knowledge production and innovation through collaboration between developing countries. These countries can leverage their diversity and jointly promote a unity that can be harnessed to foster improved global health. By continuing to follow a course emphasising South-South collaboration, Brazil can make important contributions towards cultivating southern-based knowledge production and innovation.

References

Alcorta, L., G.A. Plonski and R.C.A. (2007). "The experience of technological collaborations by Mercosur companies", Maastricht: The United Nations University, INTECH.

Asociación Redes-Biotech (2005). "Catálogo de Patentes sobre Biotecnologia no Mercosur", Documento N° 5. Available at *http://www.biotecsur.org/biblioteca-de-informacoes/inventario-e-diagnostico/5_Patentes_pt.pdf*

Assad, A., A. Corrêa, A.C. Torres and J. Henriques (2000). "Um Centro Brasileiro Argentino para a biotecnologia", *Parcerias Estrategicas* 9: 154-167.

Bialy, H. (1998). "CYTED-BT: An international biotechnology network that works", *Nature Biotechnology* 16(794): 794.

Brasil Ministério de Saúde (2005a). *Informe sobre a Reunião para Estudo de Protocolo de CooperaçãO entre os Governos de Brasil e Argentina na Área de saúde, com destaque para Medicamentos e Reativos para Diagnóstico.*

———. (2005b). *"Protocolo de intenciones entre os Ministérios da Saúde da República Federativa do Brasil e o Ministério da Saúde e Ambiente da República Argentina sobre Cooperação na Área da Saúde sobre medicamentos"*, in Roche Correa, A. P. Cooperação entre os Governos do Brasil e da Argentina na Área de Saúde.

Carullo, C. (2005). *Biotecnologia e biossegurança - Integração e oportunidades no mercosur.*

CDC (2009). "Meningococcal diseases and meningococcal vaccines", *Factsheet*, Centers for Disease Control and Prevention.

CNPq (2007). *Relatorio Institucional do CNPq, 2003-2006*. Brasilia: Conselho Nacional de Desenvolvimento Científico e Tecnológico.

Costa Vaz, A. and C.Y.A. Inoue (2007). *Funding international development: The role of emerging economies—The Brazilian case*. Ottawa: The International Development Research Centre.

Dellacha, J.M. (2003). *La biotecnología en el Mercosur: Regulación de la bioseguridad y de la propiedad intelectual*. Universidad Nacional del Litoral.

Ferrer, M., H. Thorsteinsdóttir, U. Quach, P.A. Singer and A.S. Daar (2004). "The scientific muscle of Brazil's health biotechnology", *Nature Biotechnology:* DC8-DC12.

Grogg, P. (2010). *Cuba, Brazil unite for Africa's health*. Inter Press Service.

Guimarães, J. (2000). "Centro Argentino-Brasileiro de Biotecnología: Uma experiência pioneira e exitosa de integração e cooperação regional em ciência e tecnologia", *UBA e SECYT*. Universidad, Ciencia y Tecnología en el MERCOSUR.

IBSA (2005). *India-Brazil-South Africa dialogue forum*. Cape Town: Ministerual Communique.

———. (2007). *India-Brazil-South Africa dialogue forum*. New Delhi: Ministerial Communique.

Lafer, C. (2000). "Brasil: dilemas e desafios da política externa", *Estudos Avançados* 14: 260-267.

Lage, A. (2007). "Connecting science to population health: The 'closed loop' approach", *MEDICC Review* 9(1): 48.

———. (2008). "Connecting immunology research to public health: Cuban biotechnology", *Nature Immunology* 9(2): 109-112.

Marcano González, L.F. (2006). "Science and technology framework program for Mercosur", *Interciencia* 31(11): 770-770.

Mathias, M. (2002). "Inserção internacional: Três temas na agenda brasileira", *São Paulo em Perspectiva* 16: 27-35.

Ministry of Science and Technology (2010). "Brasil: Dispêndio nacional em ciência e tecnologia (C&T)(1), 2000-2008", in G.O. Brazil (ed.) Brasilia.

Mokoena, R. (2007). "South-South co-operation: The case for IBSA", *South African Journal of International Affairs* 14(2): 125-145.

Narváez-Berthelemot, N., J.M. Russell and L. Velho (1999). "Scientific collaboration of the Mercosur countries as an indicator of Latin American regional activity", *Research Evaluation* 8(2): 83-90.

Niosi, J. and S.E. Reid (2007). "Biotechnology and nanotechnology: Science-based enabling technologies as windows of opportunity for LDCs?", *World Development* 35(3): 426-438.

OECD (2009). *Globalisations and the emerging economies: Brazil, Russia, India, Indonesia, China and South Africa.* Paris: Organisation for Economic Co-operation and Development.

Petherick, A. (2010). "High hopes for Brazilian science", *Nature* 465: 674-675.

Ponce, C., E. Ponce, E. Vinelli, A. Montoya, V. de Aguilar, A. Gonzalez, B. Zingales, R. Rangel-Aldao, M.J. Levin, J. Esfandiari, E.S. Umezawa, A.O. Luquetti and J.F. da Silveira (2005). "Validation of a rapid and reliable test for diagnosis of chagas' disease by detection of Trypanosoma cruzi-specific antibodies in blood of donors and patients in Central America", *J Clin Microbiol* 43(10): 5065-5068.

Puri, L. (2007). "IBSA: An emerging trinity in the new geography of international trade", *United Nations Conference on Trade and Development*. 35 (ed.). United Nations.

Quach, U., H. Thorsteinsdóttir, J. Renihan, A. Bhatt, Z. Aesch, P. Singer and A. Daar (2006). "Biotechnology patenting takes off in developing countries", *International Journal of Biotechnology* 8(1): 43-59.

Rezaie, R., S.E. Frew, S.M. Sammut, M.R. Maliakkal, A.S. Daar and P.A. Singer (2008). "Brazilian health biotech--fostering crosstalk between public and private sectors", *Nature Biotechnology* 26(6): 627-644.

Saraiva, M.G. (2007). "As estratégias de cooperação Sul-Sul nos marcos da política externa brasileira de 1993 a 2007", *Rev. Bras. Polít. Int.* 50(2): 42-59.

Sáenz, T.W. (2005). "Biotechnology for medical applications: The Cuban experience", *Science Technology & Society* 10(2): 225.

———. (2008). "The path to innovation: The Cuban experience", *Journal of Technology Management and Sustainable Development* 7(3): 205-221.

Soares de Lima, M.R. (2005). "A política externa brasileira e os desafios da cooperação Sul-Sul", *Rev. Bras. Polìt. Int.* 48(1).

Souza-Paula, M.C., I.T.G. Alves and C. Roitman (2005). *Centro Brasil Argentina de Biotecnologia - CABBIO.*

The World Bank (2007). *World development indicators database: Gross domestic product 2007.*

Thorsteinsdóttir, H. (2007). "The role of the health system in health biotechnology in developing countries", *Technology Analysis & Strategic Management* 19(5): 659-675.

Thorsteinsdóttir, H., A.S. Daar, P.A. Singer and E. Archambault (2006). "Health biotechnology publishing takes-off in developing countries", *International Journal of Biotechnology* 8(1): 23-42.

Thorsteinsdóttir, H., T.W. Saenz, U. Quach, A.S. Daar and P.A. Singer (2004). "Cuba-innovation through synergy", *Nature Biotechnology* 22: DC19-24.

Thorsteinsdóttir, H., P.A. Singer and A.S. Daar (2007). "Innovation cultures in developing countries", *Comparative Technology Transfer and Society* 5(2): 178-201.

Umezawa, E.S., S.F. Bastos, J.R. Coura, M.J. Levin, A. Gonzalez, R. Rangel-Aldao, B. Zingales, A.O. Luquetti and J.F. da Silveira (2003). "An improved serodiagnostic test for Chagas' disease employing a mixture of Trypanosoma cruzi recombinant antigens", *Transfusion* 43(1): 91-97.

Velho, L.M.L.S. (1997). "Cooperação em ciência e tecnologia no Mercosul", *Relatório de Pesquisa*: 1-50. Brasília: OEA/MCT. (Journal Article).

WHO (2007a). *Meningitis in Africa: Hundreds of thousands vaccinated.* World Health Organization.

————. (2007b). *WHO coordinates response to meningitis outbreaks in four African countries.* Geneva: World Health Organization. Available at *http://www.who.int/mediacentre/news/notes/2007/np12/en/index.html*

5 Beyond the Great Wall: China's South-South Collaborations in Health Biotechnology

AUTHORS: *Wen Ke, Zhang Jiuchun, Li Lexuan, Chen Guang, Christina C. Melon, Halla Thorsteinsdóttir*

5.1 Introduction

In the late 1970s China started to transform itself and its image and opened up to the rest of the world. This transformation is in stark contrast to its foreign relations during the Cultural Revolution (from 1966 to 1976) when political, economic, and cultural interactions with the western world were effectively shut off. China has, from the late 1970s, actively established diplomatic relations with the western world and sought new economic relationships in the global community. It is the world's most populous nation with over 1.3 billion people in 2007 (UNDP, 2010) and surpasses Japan by the end of 2010 to become the world's second largest economy in terms of GDP (UNESCO, 2010). China, an emerging economy, is classified by the World Bank as a lower/middle-income country (OECD, 2009; The World Bank, 2007). It is also a country that has experienced steady and impressive economic growth during the last decades with around 10 per cent annual GDP growth rates in the last few years (The World Bank, 2010). The large population and increasing economic growth has led to a rapidly growing middle-class population with expanded consumer demand. It is forecast that in 2025, the urban middle classes of China will have reached 612 million and that their spending will have increased to more than US $2.3 trillion per year (Farrell *et al.*, 2006).

China has aspirations to become a world leader in several science-intensive fields and, as a result, its investment in science and technology has increased massively. The government has allocated considerable resources for building up scientific infrastructure in the country and has set up ambitious programmes to train human resources in scientific fields as well as to attract Chinese nationals who have trained and worked abroad back to the country. China's share of the world R&D expenditure (GERD) increased from 5 per cent in 2002 to 9.2 per cent in 2007 (UNESCO, 2009). It has also put into motion several measures to encourage private sector development in science-intensive fields (Frew *et al.*, 2008).

China's openness has now extended into scientific fields and its participation in the international scientific community has grown by leaps and bounds. It has become one of the lead contributors to scientific publications in international peer-reviewed journals; in 1998 it was in ninth place in terms of number of papers in the Science Citation Index, but in 2008 it had reached second place, just after the United States (OECD, 2010). We can also see the same pattern in the health biotechnology field and, as discussed in Chapter 2, by 2009 China had become the second largest contributor to the field globally. The openness to the rest of the world is not only demonstrated in increasing publications in international peer-reviewed journals, but also in growing collaboration with experts in other countries (OECD, 2010). In the health biotechnology field our previous research has shown that China's researchers and entrepreneurs are now actively engaging in joint projects with experts worldwide (Thorsteinsdóttir *et al.*, 2007; Zhenzhen *et al.*, 2004).

While China's scientific ties with countries such as the United States and Japan have increased considerably, China is also seeking collaboration with other developing countries (not just developed ones) as its partners for international relations. Even though there has been an increasing international focus on China's South-South collaboration, for instance, reflected in the dialogue about China's collaboration with India and with African countries, research is lacking on its South-South collaboration in science-intensive fields such as health biotechnology. To address this gap we have carried out a study on China's South-South collaboration in health biotechnology, with a specific focus on China's collaboration with India, Thailand and Cuba (Box 5.1). We have assessed the policy environment of China's South-South collaborations, the extent and geographical pattern of its health biotechnology collaboration, the potential and reasons for the collaboration, and the challenges and impacts associated with China's South-South collaborations to date.

BOX 5.1

Case-study research on China's collaborations: Methodology

To gain an understanding of collaborations between China and its Southern partners, we carried out case-study research on both research and entrepreneurial collaborations in the health biotechnology field. We focussed particularly on the partnerships between China and India; China and Thailand; and, to a lesser extent, between China and Cuba. We considered each case to be the bi-national collaborations. This design allowed us to study China's collaboration with another emerging economy, India, and two smaller countries, Cuba and Thailand. China and India, Asia's two population giants, share many characteristics and have put a political emphasis on their collaboration for a few decades. As a part of that they have signed a number of technological cooperation agreements. Both countries have considerable strengths in health biotechnology and large markets for health biotechnology products. Thailand and Cuba have both emphasised the development of their biotechnology sectors since the early 1980s.Thailand is a neighbouring country to China and therefore may share health concerns that cross their boundaries or have populations with similar genetics. We also included a cursory examination of China-Cuba collaboration as a part of this study because of the emphasis the two countries have placed on their entrepreneurial collaboration in the health biotechnology field.

The case studies relied on multiple sources of data (see description of the methodology in Chapter 1), including scientometric analysis of co-publications, a survey of firms about their collaborations, and interviews with 38 experts in the chosen countries (Table 5.1). To supplement these data, we used document analysis of policies and background literature, other statistics of relevance to the topic.

TABLE 5.1
Breakdown of number of interviewees in case-study countries

Countries	Number of interviewees
China	20
India	9
Thailand	6
Cuba	3
Total	38

5.2 Government interest and support

In its Tenth Five-Year Plan for 2001–2005, China included international S&T cooperation as a key item in its national planning agenda (Pentas, 2005). To strengthen national innovation capabilities in S&T and to speed up technology-intensive industrialisation, China has pursued the 'opening-up' strategy mentioned above involving allocating increased resources to its international collaborations (IDRC and SSTC, 1997; OECD, 2008). For example, annual funding for international collaboration from the National Natural Science Foundation of China (NSFC) was US $2.64 million in 1996 and by 2009 it had reached US $23.69 (NSFC, 2003 and 2010). However, the push for international S&T collaboration has focussed more on Northern countries with large research systems, rather than on the South (Jonkers, 2009).

Still, South-South collaboration is on the agenda of the Chinese government. Although it has been on the receiving end of several UNDP initiatives, it is increasingly an active contributor to other developing countries. This has been primarily through involvement in the Silk Road Initiative to enhance cooperation among China and neighbouring countries in Central Asia (e.g., Kazakhstan and Kyrgyzstan); and the programme for 'Promoting South-South Cooperation in the 21st Century' through capacity building, networking, and the establishment of public-private partnerships (PPPs). Furthermore, China has been actively building links with the ASEAN group of countries, which includes Indonesia, Malaysia, the Philippines, Singapore and Thailand. While this regional collaboration includes only fellow Asian countries, China's collaboration focus has been augmented with strong ties to African countries as well. It has, for example, signed bilateral agreements incorporating scientific cooperation with 11 African countries, and established a US $5 billion development fund for Africa to support cooperative projects between African nations and China (Hassan, 2007a and 2007b). This is further discussed in Chapter 9 on sub-Saharan African collaborations with China and India.

As health biotechnology is a field that China has emphasised, there is scope for China to work with other developing countries in this field to extend capacity building in other countries. For the last two decades, medical biotechnology has been singled out as an important area of Chinese S&T policies, and was one of six priority fields in the so-called '863' programme from March 1986 (Zhenzhen *et al.,* 2004). As mentioned above, China's scientific publications in international peer-reviewed journals have increased exponentially in recent years (Zhou and Leydesdorff, 2005). Its health biotechnology publications have increased, to the rate of almost fivefold from 1991 to 2002, making it the leading developing country to publish in this field (Thorsteinsdóttir *et al.,* 2006). It also has a significant private sector development in health biotechnology (Frew *et al.,* 2008). China has not drawn up a specific South-South collaboration policy or programme to promote international collaboration in the health biotechnology field. It has, however, focussed R&D activities on international cooperation on Traditional Chinese Medicine (TCM), a priority field in China (Li *et al.,* 2005). In 2006, several ministries in China jointly launched the International Traditional Chinese Medicine Program for Cooperation in Science and Technology, and set up a special fund to finance it (Embassy of the People's Republic of China in Ireland, 2008). Further, some of China's bilateral agreements, including with India, are focussed specifically on the biotechnology sector. Therefore, while China lacks a formal policy on South-South collaboration in health biotechnology, its political emphasis on developing its biotechnology sector and strengthening S&T ties with other developing countries has created a policy environment that is increasingly supportive.

5.3 The geography of China's health biotechnology collaboration

5.3.1 Mapping research collaboration

To map China's main research collaborations with developing countries in health biotechnology, we identified health biotechnology papers co-authored by researchers from China with at least one other low- and middle-income country, then used these as a proxy for collaboration. We identified papers from the Scopus database for the period between 1996 and 2009 (see Chapter 1 for details of the methodology). We identified 973 papers that China has co-authored with other developing countries for the period studied. Compared to other low- and middle-income countries, China is in second place and lags slightly behind the leader in South-South in the health biotechnology field, Brazil. Its South-South collaboration seems to have increased since 2002, when it was around 40 papers per year and had reached around 170 papers per year in 2008.

Our findings suggest that China's main research collaborations are with other countries in Asia (Figure 5.1)—six of the top 10 countries come from this region, with India and Thailand being the top two co-authoring countries. China also appears to collaborate with emerging economies outside of Asia. Brazil is the third most common collaborator and South Africa is fifth.

FIGURE 5.1

The main developing countries China collaborates with in South-South health biotechnology research, based on the number of co-authored papers between 1996 and 2009

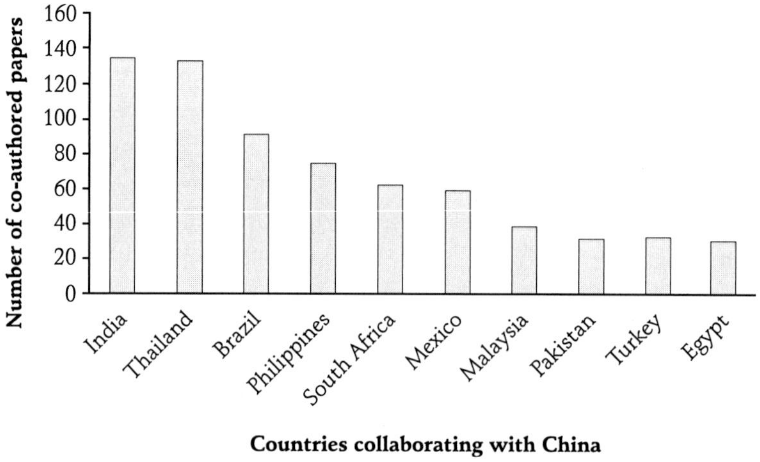

Source: Authors' presentation of data compiled by Science-Metrix using Scopus.

We looked at the subfields of China's South-South health biotechnology collaboration. Not surprisingly we observed the same pattern as generally for South-South collaboration discussed in Chapter 2. 'Genetics and Heredity' was the most common subfield of China's South-South collaboration, 'Biochemistry and Molecular Biology' was in second place, and 'Microbiology' was the third most common subfield. Subfields that are particularly closely aligned with health needs that exist in developing countries but not in Northern counties, i.e., 'Parasitology' and 'Tropical Medicine' were not common subfields of China's South-South collaboration with 'Parasitology' ranking in 10th place and 'Tropical Medicine' ranking in 12th place.

We compared the extent to which China collaborates with developed countries in the health biotechnology field. With high-income countries, China had 17,852 co-authored papers for the period between 1996 and 2009. It had many more co-publications with high-income countries than the other countries in our study. China has over three times more such papers than Brazil and more than four times the number of co-authored papers than India has with the North. It is obvious that most of China's partnerships in the health biotechnology field are with high-income countries. The United States is by far China's most common Northern partner (10,078 co-authored papers), which is not surprising considering the global role the United States plays in health biotechnology. Japan is then in second place (2,514 co-authored papers) which is likely to reflect its geographical proximity to China. The United Kingdom was in third place (1,901 co-authored papers).

5.3.2 Mapping entrepreneurial collaboration

In order to identify China's main partners in entrepreneurial collaborations, we administered a survey to all the Chinese firms we could identify that were involved in health biotechnology activities, as well as firms in the main developing countries active in this field: Brazil, Cuba, Egypt, India and South Africa (see Chapter 1 for details of the methodology). We considered entrepreneurial collaboration to be joint work involving firms in activities ranging from R&D to the commercialisation of products. From these data, we then graphed the countries with which China had at least three collaborations (Figure 5.2). China's South-South collaboration levels are surprisingly low, with only 11 per cent of firms in China reporting collaboration with other firms in developing countries in this field. As a result China has three or more collaborations with only four partners. The emerging economies, India and Brazil, are the main countries with which China collaborates, but Chinese entrepreneurs also have partnered with Egypt and Cuba. By far, the main collaborative activities reported were marketing and distribution.

FIGURE 5.2

The main developing countries China collaborates with in South-South entrepreneurial collaborations in health biotechnology, based on a survey of firms

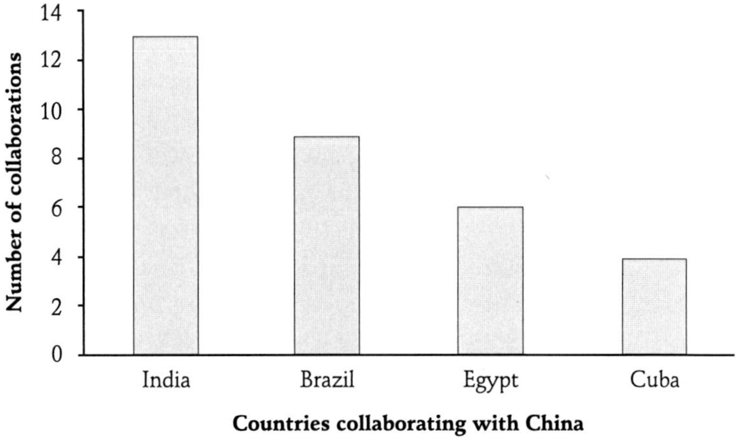

Source: Authors' presentation of their own survey data.

To gain a deeper understanding of collaborations between China and its Southern partners, we focussed particularly on examining the collaborations between China and India; China and Thailand; and, to a lesser extent, between China and Cuba. As India was the only country to rank among China's top two partners for both research and entrepreneurial collaborations, it was of interest to learn what drives these collaborations. The governments of China and India have placed a strong emphasis on their technological collaboration. Their diplomatic relations began in 1950, shortly after the People's Republic of China was founded. Between 1988 and 2006, the leaders of China and India signed a total of 15 bilateral cooperation agreements focussed on technological cooperation, and singling out biotechnology, genomics, and TCM as fields of key interest. In Chapter 7 we focus particularly on India's South-South health biotechnology collaboration and map and analyse further its collaborations with developing countries.

We also wanted to choose a country in the region that was not considered to be as strong in health biotechnology as China, but still had experience in collaborating with China. The Philippines and Thailand were the main candidates, as both have engaged in research collaboration with China. We chose to focus on the collaboration with Thailand because its proximity to China, compared to the Philippines, may increase the likelihood of sharing similar health concerns. Also, China has a bilateral technological agreement with Thailand that dates back to 1978, and in 2007, a Sino-Thai strategic cooperation action plan was formed to further strengthen cooperation in the field of technology. Lastly, Thailand has targeted the biotechnology sector since the early 1980s, and established research infrastructure and policies to promote it. In 1983 the government established a key biotechnology centre, what is now called BIOTEC (Pathumthani, Thailand) but previously was known as the 'National Centre for Genetic Engineering and Biotechnology.' It operates heavily in agricultural and food biotechnology fields but includes an emphasis on medical biotechnology as well. The government developed a specific policy to promote biotechnology through Thailand's 'National Biotechnology Policy Framework' (2004–2009) (NCGEB, 2005). Its strategic goals include to build Thailand into a key health care sector in Asia and to establish a network of highly qualified biotechnology human resources. Thailand's main collaborators in the health biotechnology fields are the same as China's; the United States (785 papers), Japan (668 papers), and the United Kingdom (283 papers). China ranks fifth (133 papers) as Thailand's key collaborator, and the next developing country collaborator is India in 11th place (44 papers).

To get fuller insight into China's industry collaboration in the health biotechnology field, we supplemented our case-study research with a look at China's entrepreneurial collaborations with Cuba. Our mapping exercise identified Cuba as one of China's main entrepreneurial collaborator in the field. China and Cuba signed a bilateral agreement in 2004 that is exclusively focussed on biotechnology collaboration. Also, as discussed in Chapter 4, Cuba has placed significant emphasis on the development of biotechnology products, and has a relatively strong innovation record in the field (Thorsteinsdóttir *et al.,* 2004b). It also has frequent South-South collaborations as described in Chapter 4.

5.4 Collaboration in research activities

The China-India and China-Thailand research collaborations covered diverse topics in health biotechnology, including genomics, immunology, and gastroenterology. Almost all the research collaborations studied here included only bi-national collaboration, but the health biotechnology collaborations in general between these countries frequently involve Northern partners as well. The research topics include studying hereditary structures of people of South East Asia; mitochondrial DNA; basic research on immunology; Angiostrongylus cantonensis, a cerebral spinal disease caused by a parasite; and Chinese herbal medicine. It appeared from our interviews that the initial push for the collaboration originated mostly through the collaboration partners themselves, rather than external bodies such as governments or international organisations. Additionally, we found that Chinese and non-Chinese partners seem to initiate the collaborations equally.

5.4.1 Reasons for the collaboration

The reasons behind the research collaboration cases we studied seem to be diverse, and in many cases several reasons were found to motivate a particular collaboration. Some of the more prominent reasons included:

1. *To gain access to expertise/technologies:* As science in general becomes more specialised, there is an increased need for researchers to work together to provide the different types of expertise needed for any one research project (Bozeman and Corley, 2004; Gibbons, 1994; Katz and Martin, 1997). Gaining access to expertise was an important reason for the collaborations we studied here. China has, for example, been building capacity in genomics, and was the only developing country to take part in the Human Genome Project (Lander *et al.,* 2001). The main driver for the collaboration on the mitochondrial DNA of Indians—between the Kunming Institute of Zoology (Kunming) of the Chinese Academy of Sciences, the Sanjay

Gandhi Postgraduate Institute of Medical Sciences (Lucknow), and the North Bengal University (Siliguri)—was to gain access to Chinese sequencing infrastructure and expertise (Box 5.2). Further,

BOX 5.2

The genomics ties of China and India

As the only developing nation to have contributed to the Human Genome Project, China now has a strong reputation in the field of genomics. As a result, China is a sought-after partner in genomics research and publishes more South-South collaborative papers in the 'Genetics and Heredity' subfield of health biotechnology than any other developing country.

One such project involved sequencing the mitochondrial DNA (mtDNA) of Indian samples. This was done to compare them to the phylogeny of Eurasian samples and allow for phylo-geographic screening in the future. However, applied to the medical context, findings from this work could also be applied to systemic studies of mitochondrial diseases, thus bringing about potential health gains for the population being studied as well. As described by the authors of one of the three published papers stemming from this work, 'In patients with South Asian matrilineal ancestries, at least a basal outline of the total mtDNA phylogeny in this subcontinent is indispensible' (Palanichamy *et al.,* 2004). For a population where such information is currently sparse, even such an overview can begin to open the doors to the potential of future screening for mitochondrial diseases.

In order to carry out the project, Indian researchers from the Sanjay Gandhi Institute of Medical Sciences (Lucknow) joined efforts with the Kunming Institute of Zoology (Kunming) at the Chinese Academy of Sciences. While China provided the sequencing equipment and technical expertise, India contributed the much needed mtDNA samples, and both parties provided primers, methods, and complementary expertise. Beyond the potential health and other applications of the project findings, the work has also benefitted India in developing capacity in the field of genomics. It has, for example, led to the formal training of a postgraduate fellow in this field and allowed India to 'take advantage of the sequencing technology on mtDNA of the [Chinese] side,' while giving the Chinese partner access to the necessary mitochondrial samples for sequencing.

while capacity in China for health biotechnology is significant, publishing 10 times more in international peer-reviewed journals than Thailand, Thai researchers have pockets of expertise that are essential for Chinese researchers. This appeared, for example, to be the case in the collaboration on Beta-Thalassemia syndromes between Guangxi Medical University of China (Nanning) and the Thalassemia Research Center, Institute of Science and Technology for Research and Development, Mahidol University of Thailand (Nakhonpathom), where the latter group provided years of experience in studying the syndrome to the collaboration. Instead of building expertise and infrastructure in a wide range of health research subfields, it can be valuable to focus on specific niche areas and collaborate with neighbouring countries or other global players to gain access to complementary knowledge in these areas.

2. *To acquire access to clinical samples:* The success of projects in health biotechnology can depend on access to suitable clinical samples in sufficient quantities. Research at the Zhejiang Academy of Medical Sciences (Hangzhou) in China on the parasite Angiostrongylus cantonensis, which was causing an increasing prevalence of eosinophilic meningitis in China, required samples from Thailand, where there is traditionally a higher incidence of the disease, and collaboration with the Department of Parasitology, Faculty of Medicine, Chiang Mai University (Chiang Mai, Thailand) to acquire enough samples of the parasite to do the research. Likewise, Chinese researchers at Dalian Medical University (Dalian, China) needed collaboration with India when they wanted to study the immunohistochemical distribution of Ley oligosaccharides and their role in reproductive health in rhesus monkeys. As a Chinese scholar expressed: 'Without the cooperation with India, we wouldn't be able to do the research for lack of samples on rhesus monkeys.' Even though China has a vast population, with diverse diseases and several monkey colonies to carry out biomedical research, there are cases when samples are needed from other countries (Hao, 2007).

3. *Shared health concerns:* Health problems do not respect national borders, and it can be of paramount importance for neighbouring

> countries to collaborate in order to deal successfully with them. This applied to the case of Angiostrongylus cantonensis discussed above. When the incidence of eosinophilic meningitis was on the rise in China, researchers sought to collaborate with Thailand, where local researchers had developed expertise on this topic. As a policymaker in Thailand stated: 'Emerging diseases is a domain to be developed with priority, such as avian flu, dengue fever and malaria. [These are conditions] which developed countries care less about.' The message from the interviews in this study was that developing countries approached the North for capacity building, but needed to work with the South to address their health problems. In the words of a Thai policymaker, 'From the perspective of technology, we can learn a lot from South-North cooperation; while in South-South cooperation both sides have more motives and initiatives, and can be complementary to each other, which makes more sense in the long term.'

While equal contribution by the different partners is generally considered to be a preferred collaboration model, it is sometimes appropriate to employ unequal collaboration to develop the capacity of the weaker partner. As one Chinese interviewee noted,

> Our counterpart provided us with some data and materials, which were extremely helpful to our research, especially those materials on some rare diseases. We basically provided them with techniques. After cooperating with us, they could also master these techniques and carry out such kind of research.

This illustrates not only that the motives for collaboration are multifold, as outlined above, but also that, despite the differences in roles, the contributions of both parties seem invaluable to the success of the project and capacity-building efforts.

5.4.2 Challenges to the collaboration

There were several challenges, both in the China-India and China-Thailand collaborations, which posed a threat to their successes. They included:

1. *Funding:* Neither China, India, nor Thailand, have dedicated funds to support South-South collaborations in the health

biotechnology field, and raising funds for research turned out to be a major challenge for their research collaboration. This was a theme reported in most of our interviews. Working with other developing countries, particularly smaller nations, or those where biotechnology is not a specific focus of the government, made it especially difficult to raise enough money for the South-South research collaboration. As one Chinese researcher stated: 'Without support of funds, there won't be serious and close cooperation, but only [a] loose one.' Lack of funding appears to direct the international collaboration more to collaboration with Northern countries. One of China's research collaborations with Thailand was, for instance, carried out in collaboration with Japan, in order to get funds for the joint research.

2. *Transporting samples:* Some countries' regulations can restrict the movement of biologicals or other samples across borders. This is a substantial obstacle to international collaboration in general, and was named by several of our interviewees as a cause of disruption in their research. In some cases, the country with the more flexible regulations was able to send its samples to China, where the work was subsequently carried out. In other cases, researchers had to carry samples across borders themselves. This is a common concern in the field, where samples are often time sensitive and fragile to temperature fluctuations. Raising the influence of world events on the problem, one Chinese researcher stated: 'We all face this problem, which is mainly the restriction on transportation of bacteria strains after the [September 11, 2001] attack in the United States.' In our interviews with others, we have seen the challenges that tighter security at border crossings have imposed on collaborations and the movement of materials.

3. *Communication:* Several of the interviewees stated that communication was a particular challenge in their China-India and China-Thailand collaborations. The cultures differ in their communication styles and the lack of a shared language can accentuate this problem. As one Thai interviewee stated, 'Culture is the major obstacle of South-South cooperation. Researchers from different countries have different working [styles] and ideas,

> which needs coordination and tolerance. Language is another problem.'

Even those most experienced with international collaboration are not immune to the challenges of working internationally. Still, with more experience, researchers seem to develop strategies on how to handle those challenges and to learn how to work together and gain understanding of each other's expertise and working conditions. Effective communication is a valuable skill that can provide a solid foundation for further collaboration.

5.4.3 Impacts of the collaborations

When we asked interviewees to gauge the effects of their South-South collaborations, they generally felt the collaborations had positive outcomes, and there was enthusiasm expressed about the collaborations. One Indian researcher said, for example, 'Working together, China and India can make achievements far better than America does.' The collaboration can also speed up the research process. As an Indian researcher commented, for example, 'My research team is more willing to cooperate with developing countries like China, where the environment is good and people are efficient. [My collaborator] can finish in seven days the work that takes us about seven months to finish, and their equipment is also better.'

When asked to describe the outcomes of the collaboration more specifically, the typical response we heard was that it led to a publication, which reflects that the most common impact of research collaboration is the contribution of new knowledge in the field. Some of the research collaborations focussed on basic research, and apart from advancing the knowledge frontier, they were not likely to have direct applicability. As one of our Chinese interviewees stated: 'We are all pure scientific researchers, so we didn't consider the application in the future.' Not all the collaboration was confined to basic research, as some focussed on strengthening developing countries' research on shared health problems. Unfortunately, it can often take years for even a simple practical application to be realised, so the collaborations are not likely to have easily measurable impacts in the near future. More research on joint health concerns may strengthen developing countries' capacity in the long run to develop solutions to their health problems, and contribute towards improved global health.

Some of our interviewees also remarked that the collaborations resulted in training and building health biotechnology capacity. It is noteworthy that our interviewees stressed that both China and India benefitted from their collaborations in terms of capacity building, and that this was not limited to Thailand as the weaker country in health biotechnology. The collaborations we examined often involved sending graduate students and post-doctoral fellows between the participating countries to receive training in the subfields of health biotechnology, or the application of particular technologies. They could then bring this knowledge back to their home countries, making capacity building an important result of South-South health biotechnology research collaboration, as gauged by our study.

5.5 Collaboration in entrepreneurial activities

In researching China's South-South biotechnology collaborations, we also studied China's entrepreneurial partnerships through a series of interviews with representatives from firms and policymakers. In our initial survey of health biotechnology firms, we saw that Indian firms are the most common partners for China, but that collaborations between these two giants rarely involves R&D activities. Instead, they seem to focus only on marketing/distribution activities or on providing supplies; the typical scenario in the latter case involves Chinese firms providing Indian firms with active pharmaceutical ingredients. As a result, our discussion that has a stronger developmental focus is drawn from our research on China-Thailand and China-Cuba collaborations. Even though these collaborations are limited in number, they have, according to our survey on South-South firm linkages, a stronger focus on R&D activities than the China-India collaborations. From the China-Thailand and China-Cuba collaboration cases, we received the consistent message that governmental support was important for initiating and promoting the entrepreneurial collaborations. Although the governments of China and India have announced their interest in working together, signed several bilateral agreements, and singled out biotechnology as an important focus for collaboration, we have not seen these elements play a significant role so far in promoting their bilateral entrepreneurial collaboration.

5.5.1 Reasons for the collaborations

Our interviewees indicate that the main reasons for South-South collaborations in the countries we focussed on are as follows:

1. *Access to markets:* Given that the primary activities in the entrepreneurial collaborations between China and India were marketing and distribution, it is not surprising that company representatives in China indicated that their main motivation for collaborating with India was to gain access to new markets. One interviewee explains: 'Without international cooperation, we can also somehow sell in the India[n] market... However, for overall cultivation of the market and sustained development, collaboration with Indian companies will have more potential for growth in the future.' Our survey showed that Indian firms also mentioned access to markets as a key reason for their collaborations with China. These countries have large populations, so gaining reciprocal market access can be lucrative for both parties. Some Chinese interviewees explained that collaborating with local Indian firms is beneficial for gaining knowledge of local regulations and distribution networks, as well as becoming better aware of the local culture and consumers' expectations for health biotechnology products.

2. *Access to technology/expertise:* China-Cuba collaborations were partly driven by the Cuban firms' wish to access the large Chinese market, but they also were motivated by the Chinese collaborators' wish to access the technologies and expertise of Cubans in biotechnology. To facilitate these interactions, the governments of China and Cuba have established joint ventures around technology transfer from Cuba to China. Changchun Heber Biological Technology Ltd. (Chang Heber) is a joint venture formed by Changchun Biological Products Institute from China (Changchun, China), and a Cuban firm, Heber Biotec (Havana, Cuba). Cubans have transferred the technology of producing recombinant human interferon alpha–

2b to China, and built facilities to manufacture the interferon in China together with their partners there. The benefit to Cuba has been greatly expanded market access into Asia, circumventing the costs of exporting products from Cuba to China. At the same time, Chinese firms have benefitted by gaining access to knowledge in the health biotechnology field and capacity building in manufacturing biosimilars. Biotechnology Pharmaceutical Co. Ltd. (BPL) (Beijing) is another China-Cuba joint venture, this time between CIMAB SA (Havana) from Cuba, and the Chinese firms Beijing Jingyitaixiang Technology Development Co. Ltd. (Beijing), Shanxi Xinyutong Material Commerce Co. Ltd., and China International Centre (Beijing). This joint venture involves transferring technology for manufacturing monoclonal antibodies from Cuba to China, where they are now manufactured for the first time for the Chinese market. The aim for BPL is not to be confined to manufacturing already developed Cuban products, but to advance their own product portfolio by working together with Cuban partners on the development of cancer vaccines based on monoclonal antibodies.

3. *Co-research and development of traditional medicine:* The motivation behind the only China-Thailand entrepreneurial collaboration we could identify was shared interest in developing an herbal-based treatment against HIV and AIDS (Box 5.3). The product is based on original research at the Kunming Institute of Botany, Chinese Academy of Sciences (Yunnan) on the medicinal properties of Chinese plants with continued pharmacological and toxicological research in Thailand. The Thai side provided patient cases and clinical support, and together with researchers in China, carried out Phase I to III clinical trials on AIDS patients in Thailand. The Chinese firm, SH-IDEA Pharmaceutical Company (Yunnan), became involved in this process because research institutes are banned from producing drugs in Thailand. The product has received regulatory approval to be marketed in Thailand. Without involving the Chinese firm in this process, the project would not have come to fruition.

BOX 5.3

A South-South herbal approach in dealing with HIV/AIDs

The Complex SH Project is not only a good example of Sino-Thai Cooperation, but also a good example of South-South Cooperation.

— Lu Yongxiang,
Deputy Chairman, NPCC

In 1997, a Memorandum of Understanding (MOU) was signed between the Ministries of Public Health in China and Thailand, marking the formation of a new collaboration between these two neighbours. As a part of this collaboration, they started joint development of an herbal remedy rooted in the principles of traditional Chinese medicine that would be effective against HIV/AIDS. Now, the product—Complex SH—is the first herbal anti-HIV drug to have undergone Phase I, II, and III clinical trials in China and Thailand, and it did so entirely with the support of the Ministry of Public Health in Thailand (Sangkitporn *et al.,* 2005).

The official partnership was established between the Department of Medical Science within Thailand's Ministry of Public Health, and the Kunming Institute of Botany at the Chinese Academy of Sciences (CAS). Phytochemist Professor Luo Shide at the Kunming Institute in the late 1990s started to perform *ex vivo* experiments testing effectiveness, and analysing the pharmacological and toxicological properties of his compounds. Later, these tests were performed again in Thailand, prior to the commencement of clinical trials. As Thailand has a higher reported prevalence of HIV/AIDS than China, it became a preferred partner for China because of its government's motivation to address the growing health threat, and its larger patient base facilitated clinical trial testing. Governmental interest in the promise of this compound began specifically after Thai officials visited the lab of Professor Shide (CAS) in China. However, because research institutes are restricted from commercialising products in Thailand, the inclusion of an entrepreneurial partner, the Chinese firm SH-IDEA Pharmaceutical Company was needed.

The end result is a patented compound said to inhibit growth of HIV and kill the virus by interfering with the virus's protein hydrolysis and transcription enzymes. Increasing CD-4 cell counts, complex SH has an efficacy rate of 89 per cent when used alone, and even higher when used in conjunction with allopathic therapies, and is without the side effects that accompany a similar class of biomedical drugs.

5.5.2 Challenges to the collaborations

The chief challenges expressed by our interviewees included the following.

1. *Extensive time and costs required to move products across international borders:* Several interviewees described delays with customs, and the impact these have had on their products and the collaborations. According to them, customs officials rarely know how they should classify and handle biological products. This leads to delays in the transfer of materials, and can pose a threat to both the viability of the product and the collaborative project. Also, increased costs often result from difficulties transporting products in accordance with local government rules and regulations. For example, carbon dioxide ice ('dry ice') has been banned from China, complicating the import of some compounds. To get around this, collaborators have had to send their products with specific express service companies that have the necessary permit to import/export these materials; this results in a higher financial burden for the firms.

2. *Product registration challenges:* Further difficulties for entrepreneurial collaborations surround registering products in the different countries. Because none of the countries' regulatory agencies have agreements of mutual recognition, inefficiencies occur and efforts are duplicated when product development takes place across borders. Difficulties also arise when registering medicinal plants, as most developing countries try to emulate Northern standards for regulation, which focus on isolated and purified active compounds, and are not attuned to the development of medicines from complex plant-derived mixtures. These kinds of medicines have often been used for thousands of years, but new standards for medicines require an assessment of the individual components and their safety, rather than the effect of complex mixtures as in traditional medical practices. Given the history of their use, it may be excessive in some cases to follow all the testing procedures dictated by new drug development standards.

3. *Pressure to minimise manufacturing costs:* Chinese interviewees also raised the concern that South-South collaboration can add an

extra financial burden on firms, as there may be excessive pressure from the South to drive prices down. One entrepreneur in China said for instance: 'Companies from countries with low and medium incomes press down the price hard, while companies from developed countries feel satisfied with the ratio of price to quality and they seldom bargain.' The pressure from Southern firms to strive for low costs has led to cases in which less profit is realised by the firm that provides the drugs or technology. This can act as a disincentive for collaborating with others in the South.

5.5.3 Impacts of collaboration

China's entrepreneurial collaborations with Cuba, India and Thailand in health biotechnology provide examples of how firms in developing countries can work together to expand their markets and also to strengthen their innovative efforts. Along these lines, interviewees identified product approval and registration in a new market as being a major outcome of South-South collaboration. Firms expressed optimism that expansion into Southern markets enabled building their brands globally and their South-South collaboration was their first step towards wider global dissemination. A further impact emphasised by our interviewees is that South-South entrepreneurial collaboration can play an important role in addressing health challenges common to developing countries. This was both because of their strong focus on their joint health problems and because they could offer more affordable health solutions. An interviewee involved in the collaboration to develop an herbal compound for managing HIV mentioned 'addressing a social need' as an outcome of the collaboration between China and Thailand, emphasising that the expected outcome would be a more affordable therapeutic agent that could increase the access of poorer people to HIV/AIDS treatment options.

5.6 Conclusions

Our case-study research on China's health biotechnology collaborations with India, Thailand and Cuba has identified several findings that illustrate the potential for South-South collaboration and shed light on approaches to strengthen it. The main messages from the research are:

China is steadily engaging in more South-South collaborations, but its traditional focus of partnering with the North still prevails. In the past, both researchers and firms in China have regarded collaborations with the North as being more worthwhile than those with other developing countries. Regarding their Northern partners as more advanced scientifically, they have perceived collaborations with them as being a better way to gain knowledge and advance their own scientific standing. Collaborations with developed countries have also been preferred for financial reasons, as funding has often flowed from the North to the South in these arrangements. Although there is a growing recognition that Southern partnerships have a lot to offer China as well, perceptions are slow to change, and so is the shift towards South-South linkages.

Developing countries can work together to better address local health needs that may otherwise be overlooked by developed countries. As we have seen through the examples of China's partnership focussed on Vibrio parahaemolyticus, topics and approaches undervalued by the North are taking centre stage in South-South collaborations. Developing countries are more likely to share similar research interests and health threats with each other than with developed countries, and South-South collaborations are therefore an avenue through which developing countries can effectively work on their priorities.

South-South partnerships can help preserve and advance traditional knowledge collaboratively. In the China-Thailand collaboration to develop an herbal treatment for HIV patients, a new formulation was created using principles of traditional Chinese medicine. In this particular case, a patent was generated which is now held by the Southern researcher who first developed and tested the compound. In general traditional knowledge and biodiversity are great resources of developing countries which they can jointly harness.

Formal policies and agreements are not sufficient in and of themselves to maintain collaborations. Even where bilateral agreements are in place, and official priorities favouring South-South collaboration and biotechnology advancement have been set, the impact of these initiatives is often underwhelming. As noted by our interviewees, other barriers like restrictions on exports and misalignments or inefficiencies in regulatory systems can hamper the realisation of these collaborations. We saw that most of the challenges experienced by the collaborating firms were related to a misalignment between the governments' import and regulatory systems. Further, financial resources allocated to South-South collaborations were in short supply, limiting options for collaboration. For these reasons, it is imperative that governments seeking to foster new collaborations and maximise existing ones make a concerted effort to minimise or eliminate challenges, and align their systems so they can work together efficiently. While it is beneficial to have formal, government-supported initiatives in place to spark collaboration, other measures need to follow closely if these linkages are to result in measurable outcomes.

Our case studies on China's South-South collaboration in health biotechnology demonstrate that China has come a long way in the last 30 years, from being closed to international knowledge flow and innovation, to its current relatively open state. Now, China actively works with researchers and entrepreneurs in both the North and increasingly the South to reach beyond the Great Wall to advance the knowledge frontier, and promote innovation in health biotechnology. China clearly plays both a give and take role in this field, but stronger government will and more dedicated resources could help further cultivate South-South collaboration and help it realise its full potential.

References

Bozeman, B. and E. Corley (2004). "Scientists' collaboration strategies: Implications for scientific and technical human capital", *Research Policy* 33(4): 599-616.

Embassy of the People's Republic of China in Ireland (2008). *International traditional Chinese medicine program for cooperation in science and technology*.

European Commission (2010). *Community research and development information service* (CORDIS). Accessed from: *http://cordis.europa.eu/home_en.html*

Farrell, D., E. Beinhocker, U. Gersch, E. Greenberg, E. Stephenson, J. Ablett, M. Guan and J. Devan (2006). *From 'made in China' to 'sold in China': The rise of the Chinese urban consumer.* San Francisco: McKinsey Global Institute.

Frew, S.E., S.M. Sammut, A.F. Shore, J.K. Ramjist, S. Al-Bader, R. Rezaie, A.S. Daar and P.A. Singer (2008). "Chinese health biotech and the billion-patient market", *Nature Biotechnology* 26(1): 37-53.

Gibbons, M. (1994). *The new production of knowledge: The dynamics of science and research in contemporary societies*. Sage.

Hao, X. (2007) "Monkey research in China: Developing a natural resource", *Cell* 129(6): 1033-1036.

Hassan, M.H.A. (2007a) "A new dawn for science in Africa", *Science* 316(5833): 1813.

Hassan, M.H. (2007b). "Building capacity in the life sciences in the developing world", *Cell* 131(3): 433-436.

IDRC and State Science and Technology Commission (1997) *A decade of reform: Science and technology policy in China*. IDRC and the State Science and Technology Commission (People's Republic of China).

Jonkers, K. (2009) "Emerging ties: Factors underlying China's co-publication patterns with Western European and North American research systems in three molecular life science subfields", *Scientometrics* 80(3): 775-795.

Katz, J.S. and B.R. Martin (1997). "What is research collaboration?", *Research Policy* 26(1): 1-18.

Lander, E.S. *et al.* (2001). "Initial sequencing and analysis of the human genome", *Nature* 409(6822): 860-921.

Li, Z., W. Ke and C. Guang (2005). "Developing innovative capacity in China to meet health needs", *Innovation in developing countries to meet health needs*.pp.1-154.

NCGEB (2005). *National biotechnology policy framework (2004–2009)*. Patumthani: National Centre for Genetic Engineering and Biotechnology. Available at *http://www.biotec.or.th/biotechnology-en/Document/National-Biotechnology-Policy-Framework.pdf*

NSFC (2003). *Annual Report of NSFC 2002*. National Natural Science Foundation of China. Available at *http://www.nsfc.gov.cn/nsfc/cen/ndbg/2002ndbg/no04/index.htm*

————. (2010). *Annual report of NSFC 2009*. National Natural Science Foundation of China. Available at *http://www.nsfc.gov.cn/nsfc/cen/ndbg/2009ndbg/04/index.html*

OECD (2008). *OECD reviews of innovation policy: China*. OECD.

————. (2009). *Globalisations and the emerging economies: Brazil, Russia, India, Indonesia, China and South Africa*. Paris: Organisation for Economic Co-Operation and Development.

————. (2010). *Measuring innovation: A new perspective*. Paris: OECD.

Palanichamy, M.G., C. Sun, S. Agrawal, H.J. Bandelt, Q.P. Kong, F. Khan, C.Y. Wang, T.K. Chaudhuri, V. Palla and Y.P. Zhang (2004). "Phylogeny of mitochondrial DNA macrohaplogroup N in India, based on complete sequencing: Implications for the peopling of South Asia", *American Journal of Human Genetics* 75(6): 966-978.

Pentas, P. (2005). *National high-tech research and development program 863: Introduction, objectives, and organisation.* Available at *http://www.ppentas.com/thesis/National_RD_Programm_863.pdf*

Sangkitporn, S., L. Shide, V. Klinbuayaem, P. Leenasirimakul, N. Wirayutwatthana, P. Leechanachai, S. Dettrairat, W. Kunachiwa and V. Thamlikitkul (2005). "Efficacy and safety of zidovudine and zalcitabine combined with a combination of herbs in the treatment of HIV-infected Thai patients", *Southeast Asian Journal of Tropical Medicine and Public Health* 36(3): 704.

The World Bank (2007). *World development indicators database: Gross domestic product 2007.*

————. (2010). *World development indicators database: GDP growth (annual %).*

Thorsteinsdóttir, H., A.S. Daar, P.A. Singer and E. Archambault (2006). "Health biotechnology publishing takes-off in developing countries", *International Journal of Biotechnology* 8(1): 23-42.

Thorsteinsdóttir, H., T.W. Saenz, U. Quach, A.S. Daar and P.A. Singer (2004). "Cuba-innovation through synergy", *Nature Biotechnology* 22: DC19-24.

Thorsteinsdóttir, H., P.A. Singer and A.S. Daar (2007). "Innovation cultures in developing countries", *Comparative Technology Transfer and Society* 5(2): 178-201.

UNDP (2010). *Human development report 2010—20th anniversary: The real wealth of nations: Pathways to human development.* New York: United Nations Development Programme.

UNESCO (2009). *UNESCO Institute for Statistics fact sheet: A global perspective on R&D.*

———. (2010). *UNESCO science report 2010: The current status of science around the world.* Paris: United Nations Educational, Scientific and Cultural Organization.

Zhenzhen, L., Z. Jiuchun, W. Ke, H. Thorsteinsdóttir, U. Quach, P.A. Singer and A.S. Daar (2004). "Health biotechnology in China: Reawakening of a giant', *Nature Biotechnology* 22: DC13-DC18. Supplement.

Zhou, P. and L. Leydesdorff (2005). "The emergence of China as a leading nation in science", *Research Policy* 35(1): 83-104.

6 Learning through Collaborations: Egypt's South-South Health Biotechnology Collaboration

AUTHORS: *Magdy Madkour, Sahar Aly, Nefertiti El-Nikhely, Marwa Gamal Elwakil, Heba Maram, Halla Thorsteinsdóttir*

6.1 Introduction

Egypt is located at the hub of different seaways, connecting both the Middle East and Africa. It borders diverse geographic and cultural corners of the world, and is an Arabic country firmly placed within the African continent. In the late 1900s, Egypt had suffered from geopolitical conflicts and economic ups and downs. It is a lower/middle-income country that has experienced an above average GDP growth rate in recent years or around 7 per cent, in the 2006–2008 period (The World Bank, 2010a). As Egypt is the largest country in North Africa, in terms of geographic size and population, and has a comparatively well-developed science and technology infrastructure, nearby countries, both African and Middle Eastern, place demands on it to share its capacity. Still Egypt does not prioritise funding towards R&D, and GERD, as a percentage of GDP, is around 0.25 per cent (The World Bank, 2010b). As a result, Egypt falls within the group of countries that invest the lowest of their GDP for R&D (UNESCO, 2009). Resources for research are in short supply in Egypt, which does not allow for rapid scientific and technological development.

Egypt is relatively new to the health biotechnology field and it was not until the mid-1990s that the country started to focus on biotechnology to the extent of developing detailed policies (Abdelgafar *et al.,* 2004). Health biotechnology is a small subfield of Egypt's pharmaceutical sector. Even its pharmaceutical sector is small and only 21 pharmaceutical firms are amongst the 800 top industrial firms in Egypt (IMC, 2008). Our previous research on health biotechnology innovation in developing countries showed Egypt to be weaker than the other countries we studied (Brazil, China, Cuba, India and South Africa) both in terms of number of publications and patents in the field (Quach *et al.,* 2006; Thorsteinsdóttir *et al.,* 2006). Compared to some of the leading developing countries in this field, Egypt is more likely to be engaged in collaboration to develop local capacity in health biotechnology, rather than collaboration to improve capacity among its partners. By including Egypt in this study, we have thus an opportunity both to improve the geographic coverage of our study on South-South collaboration to the Middle Eastern and North African region, and also an opportunity to study South-South collaboration from the perspective of a country that needs to gain assistance from the leading developing countries in the health biotechnology field. Thus, it is of interest to examine Egypt's participation in South-South collaborations in health biotechnology, explore its levels of research *versus* entrepreneurial collaboration, the motivations for the collaborations, identify the main challenges, and examine the impact of the collaborations to date (Box 6.1).

6.2 Government interest and support

The Egyptian government has signalled that it values South-South collaboration. It started relatively early on to adopt a strategy to meet development goals through South-South cooperation and established the Egyptian Fund for Technical Cooperation with Africa in 1981 (ESIS, 2006). The fund has been active in 30 countries in Africa, working on diverse projects in health, agriculture, water resources, and education (INSouth, 2009). Egypt was also a founding member of the Organisation of African Unity in 1963, now replaced by the African Union. It has been a member of COMESA since 1998, whose vision is to become a regional economic community that is fully integrated and internationally

BOX 6.1

Case-study research on Egypt's collaborations: Methodology

To gain a better understanding of health biotechnology collaborations between Egypt and other developing countries, we carried out case-study research on Egypt's South-South collaborations, including both research collaborations and entrepreneurial collaborations. We focussed in particular on Egypt's collaborations with China and Jordan. We studied a total of two cases and defined each case as the bilateral collaboration between Egypt and China *versus* Egypt and Jordan. China is one of the leading developing countries in the health biotechnology field and publishes extensively in the field. In comparison Jordan, like Egypt, is a relatively recent participant in the health biotechnology field. This research therefore allows us to examine and compare Egypt's collaboration with a relatively strong country in the health biotechnology field *versus* its collaboration with a country that is at a more equal level. Our previous research on health biotechnology innovation in Egypt identified collaboration with China to have been important for Egypt's entrance into the health biotechnology field (Abdelgafar *et al.,* 2004). In the research reported here, we have an opportunity to examine Egypt's collaboration with China in more detail.

The research in this chapter relied on multiple sources of data as described in Chapter 1, including scientometric analysis of co-publications, a survey of firms' South-South collaboration, and interviews with a total of 78 interviewees carried out in Egypt, China and Jordan (Table 6.1).

TABLE 6.1

Breakdown of number of interviewees in case-study countries

Countries	Number of interviewees
Egypt	34
China	16
Jordan	28
Total	78

competitive (COMESA, 2010). COMESA traces its history to the mid-1960s, with African countries expressing their pan-African solidarity following independence, and exploring alternatives to trading with former colonial powers in the north. Egypt was a founding member of the Arab League in 1945, and remains active in institutions such as the Arab League Educational, Cultural and Scientific Organization (ALECSO), which fosters political, economic, and scientific programmes designed to promote the interests of Arabic countries. Furthermore, Egypt is a member of the Greater Arab Free Trade Area (GAFTA), established in 2005, whose mission is not confined to trade but also focusses on research and technological cooperation and aims to be a forum to address intellectual property issues (Abedini and Peridy, 2008). Egypt's South-South collaboration, therefore, reflects its memberships in diverse groups of countries. On one hand, it has collaboration with sub-Saharan African countries which reflects its location on the African continent. On the other hand, it has collaboration with Arab countries mostly in North Africa and the Middle East, reflecting Egypt's Arabic cultural heritage.

According to the Industrial Modernization Centre (IMC) in Egypt, the Egyptian government increased its funding to international projects in recent years (ADE/DOL, 2004; WHO, 2005). The government is also placing greater emphasis on international R&D projects. In 2006, only US $2 million were allocated to these types of projects by the Egyptian government. The amount increased to US $5 million in 2009 (ADE/DOL, 2004). In addition, Egypt recently opened a new US $10 million centre for transferring technology and promoting South-South cooperation in science, manufacturing, technology, and industrial innovation, based in Cairo (Sawahel, 2008). The centre is co-founded by the African Union, UN Industrial Development Organization (UNIDO), and Egypt's Ministry of Trade and Industry.

Egypt has signed some bilateral agreements focussed specifically on science and technology cooperation. It signed one such agreement with China in 2006, in the aftermath of China's Africa Forum (MFA, 2008; Sawahel, 2006). The two countries aimed to set up a common fund for their joint S&T activities. As a part of this agreement, they chose health and traditional medicine as areas of particular importance, and there seems to be a wish for Egypt to learn from China's experience in these research areas. Their collaboration plans were further strengthened during the Second Ministerial Meeting of the Asian-Middle East Dialogue, held in Sharm El-Sheikh, Egypt, in April 2008. The action plan that resulted included collaborative workshops on natural disasters and communicable diseases, and focus on transfer of technology. In recent years, Egypt has also signed agreements with countries, such as Algeria, Jordan and Sudan, which include plans for cooperation in health, pharmaceutical, and biotechnology research (Boumedjout, 2008; Khader, 2008; Sawahel, 2007b). Furthermore, Egypt and Kazakhstan developed a plan in 2007 to create a joint council for S&T research which would promote cooperation in fields such as pharmaceutical and biotechnology research (Sawahel, 2007a). In these latter projects, the onus seems to be on Egypt to help these countries build capacity in these research areas.

From this, it seems clear that there is a desire for South-South collaboration on the part of the Egyptian government. Its wish to collaborate is based on a foundation built years ago in its Pan-African and Pan-Arabian collaborations, mostly involving political and economic ties. In recent years, there has been an increased emphasis on collaborations in S&T including health biotechnology cooperation. It will be of interest to examine where Egypt's main South-South ties lie in this field, and what the potential is for these collaborations.

6.3 Geography of collaboration in Egypt's health biotechnology

6.3.1 Mapping research collaboration

The mapping of Egypt's research collaboration indicates that Egypt's participation in South-South collaboration in health biotechnology is low. To examine Egypt's research collaboration, we retrieved papers that Egypt has co-authored with other low/middle-income countries from the Scopus database for the period between 1997 to 2009 (see Chapter 1 for details of the methodology). We identified only 165 papers that Egypt has co-authored with other developing countries for the period studied. Since 2006, Egypt's South-South co-authored papers seem to have increased; it published only 20 South-South co-authored papers from 1998–2001, but over 90 such papers for 2006–2009. China is Egypt's main partner in

South-South research collaboration, but it only published 30 papers with China during the 13-year period studied (Figure 6.1).

Geographic proximity does not seem to be an important reason for Egypt's research collaboration in health biotechnology. Instead, Egyptian scientists tend to co-author papers with relatively strong countries in this field, such as China, Brazil and India. The results also show that Egypt has limited Arabic and African collaborations in health biotechnology. Tunisia and Jordan are Egypt's main Arabic partners and Kenya is its main sub-Saharan African partner. Still the levels of collaboration with these countries are so low that we cannot make any predictions that Egypt is likely to have any future health biotechnology ties with them. The most common subfield of Egypt's collaboration with other developing countries was 'Genetics and Heredity'. As we discussed in Chapter 2, this is by far the most common subfield of developing countries' South-South collaboration.

In comparison to Egypt's low level in South-South collaboration in health biotechnology, its collaboration with the North seems to be thriving. Egyptian researchers co-authored over 900 health biotechnology papers with authors in high-income countries, which is around half of the total papers Egypt published in health biotechnology from 1996 to 2009. Their most common partners are researchers in the Unites States (372 papers) and Germany (137 papers). As the United States is the world's leading country in the health biotechnology field, it is the foremost collaborator of almost all developing countries. Germany seems to have particularly emphasised collaboration links with Egypt and, for example, the ministries of education in both countries organised a special initiative, the German-Egyptian Year of Science, in 2007. The two countries have supported a number of joint research projects. A noticeable trend is that these countries have started to co-fund initiatives, mostly for training in the Ministry of Higher Education in Egypt. The German 'Deutscher Akademischer Austauschdienst' (DAAD) and the Ministry of Science and Technology, for instance, co-fund training for Egyptian graduate students to study in Germany (DAAD, 2010; MHESR, 2010). This is likely to strengthen further collaborations between the two countries and may partly explain the relatively high level of co-authorship in health biotechnology between Egypt and Germany. Such financial sources allocated to training and research programmes are likely to have the effects of increasing their bilateral research collaborations.

FIGURE 6.1

The main developing countries Egypt collaborates with in South-South health biotechnology research, based on the number of co-authored papers between 1996 and 2009

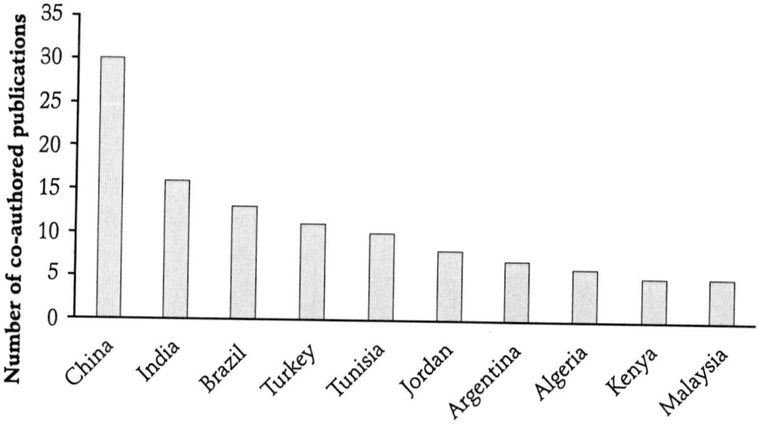

Source: Authors' presentation of data compiled by Science-Metrix using Scopus.

6.3.2 Mapping entrepreneurial collaboration

From Figure 6.2 we see that China and India are Egypt's main entrepreneurial partners in health biotechnology. As these countries are relatively strong in health biotechnology, and have a rather active private sector in this field, these are not altogether surprising results (Frew *et al.*, 2007 and 2008). What is interesting, however, is that the next-tier countries are neighbours, both the Arabic countries of Jordan and Yemen, and the African country, Sudan, suggesting that geographic proximity may play a larger role in their entrepreneurial collaboration than for their research collaboration discussed above. These countries' collaboration with Egypt may suggest that Egypt plays a gateway role between Africa and the Middle East in the entrepreneurial side of health biotechnology. By far, the main collaborative activities reported were collaborations involving distribution, but Egyptian firms also have collaborations involving the provision and use of supplies, mostly with firms in China and India where the likely scenario is that China and India provide active pharmaceutical ingredients for Egyptian production.

FIGURE 6.2

The main developing countries Egypt collaborates with in South-South entrepreneurial collaborations in health biotechnology based on a survey of firms

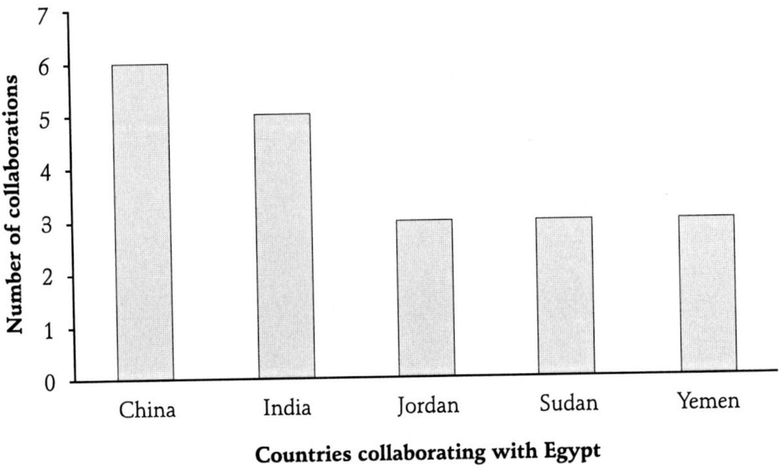

Source: Authors' presentation of data compiled by own survey.

To examine Egypt's South-South health biotechnology collaboration further, we focussed in particular on its collaborations with China, as it is Egypt's premier research and entrepreneurial partner and because there is scope for Egypt to learn from China's experience. We also chose to examine Egypt's collaboration with Jordan; a neighbouring country with similar health problems. Furthermore, Jordan has a related culture to Egypt, which may help facilitate collaborations.

Egypt has formal S&T collaboration agreements with both Jordan and with China. It also has agreements at the institutional level with several Chinese organisations. For example, the National Research Centre (Cairo, Egypt) and the Chinese Academy of Sciences (Beijing, China) have signed the so-called COMSATS (Commission on Science and Technology for Sustainable Development in the South) agreement. COMSATS is an inter-governmental organisation whose mission is to promote the use of S&T as a tool for sustainable development in Southern countries. At the 11th meeting, held in June 2008, Egyptian representatives introduced their collaborative work towards the establishment of Virtual Centres for Genetic Engineering (COMCGEB) under the COMSATS umbrella. Jordan and Egypt also have a COMSATS agreement which includes the Royal Scientific Society (Amman, Jordan) and the National Research Centre (Cairo, Egypt). The latter also has a specific agreement with the Yarmouk University (Irbid, Jordan), and there are further organisational agreements between the National Cancer Institute (Cairo, Egypt) and the King Hussein Cancer Center (Amman, Jordan). Clearly, these agreements have been signed to cultivate Egypt's collaboration with China and Jordan, and it is interesting to learn what drives these collaborations and what are their impacts, if any, so far. China's main health biotechnology collaborators are India and Thailand as was discussed in Chapter 5. Egypt therefore does not count as a key South-South collaborator for China even though it is stepping up collaboration with African countries as is discussed in Chapter 9. Jordan has only small South-South collaboration and only co-authored 40 health biotechnology papers with other developing countries for the period we studied. Its limited South-South collaboration is mostly with Middle Eastern and North African countries, where Egypt ranks in first place as its Southern collaborator.

6.4 Collaborations in research activities

As discussed above, in general, Egyptian researchers do not seem to be leaning towards South-South collaborations. They have co-authored relatively few papers in health biotechnology with researchers from other developing countries. When we interviewed researchers in Egypt, some were sceptical of the value of South-South collaboration for Egyptian health biotechnology. They argued that their collaborations were not likely to lead to scientific benefits, as both partners are at the same level of development, and suffer from the same problems. Therefore, the belief is that they cannot help each other. Instead, they advocated for collaboration with the North; to take part in the European Union's Framework programmes and collaborate with US researchers. From our interviews with researchers in Egypt, China and Jordan, it seems that many of the collaborations were formed informally, through personal contacts made at conferences and other scientific events. Personal contacts also came into play in the form of previous collaborators and supervisors. These have most frequently resulted in stronger ties with the north, as Egyptian scientists return from their studies abroad and maintain contact with their former colleagues. However, building on South-South educational exchanges might therefore also show promise as a means by which to foster continued partnerships among developing countries.

Egypt's South-South research collaboration spanned wide types of topics. They include the development of unified clinical guidelines in dealing with cancer patients between the National Cancer Institute (Cairo, Egypt) and the King Hussein Cancer Center (Amman, Jordan). There were also some cases in which Egyptian researchers took part in formal collaboration programmes, such as the genome initiative for the parasitic disease schistosomiasis, supported by the WHO. It is a large-scale mapping and sequencing initiative aimed at understanding the molecular basis of parasite metabolism and development, and determining its biological variation. The expectation is that this knowledge will be used to identify new drugs and lead to vaccine development. The initiative has several members from developed countries, as well as from developing countries such as Egypt, China, Brazil, Kenya and Vietnam.

6.4.1 Reasons for the collaboration

When we asked researchers in Egypt, China and Jordan about the main reasons for bi-national collaborations, the interviewees identified the following:

1. *Building capacity in health biotechnology:* Our respondents emphasised that Egypt needs international collaboration to build capacity in the health biotechnology field. The perception was that Egypt was not strong in the field and needed to improve its capacity through collaboration with leading developing countries in health biotechnology such as China, India and Brazil. Capacity building through South-South collaboration would make Egyptians better prepared to address local health problems, as different developing countries were likely to share similar health problems. An Egyptian researcher stated, for example: 'For Hepatitis B, therapy guidelines are different in Europe and don't go along with Egypt.

We need remedies that adapt to our local environment and meet our requirement, i.e. tailored medicine.' Others emphasised that there were smaller technological gaps between Egypt and the leading developing countries in this field, than between Egypt and developed countries, which facilitated the collaborations.

2. *Knowledge exchange around complementary strengths:* A general characteristic of the Egypt-Jordan collaborations we examined was that they were driven by the partners sharing complementary expertise. Both partners had some strengths to offer the collaboration, and by working together, they could learn from each other and increase the chances that the research project would be successful. A good example of this is the collaboration in stem cell research discussed in Box 6.2. Collaborations between the two countries in cancer genetics is another good example of complementary expertise. The National Cancer Institute (Cairo, Egypt) offers strengths in clinical expertise, supported by its work with a larger patient base, to King Hussein Cancer Center (Amman, Jordan). Jordan, on the other hand, stands to contribute its advanced experiences in patient care systems within the field of oncology. That the two countries are both Arabic and share many cultural similarities has facilitated their shared learning. An important aspect of the collaboration is the organisation of workshops and training courses. As a result of this work, both parties have benefitted substantially: more cases of cancer in Jordan are now being accurately diagnosed and managed, and Egypt has adopted approaches to health care provision based on the Jordanian model.
3. *Gaining access to equipment:* Egypt's international collaborations in health biotechnology, including the South-South ones, to a certain extent, have been focussed on gaining access to necessary equipment for the research. Through collaboration researchers can use each others' equipment and successfully complete research projects that otherwise would not be possible to conduct. Some researchers complained that it was difficult for them to identify potential collaborators that had the equipment in demand. To facilitate South-South research collaboration around sharing

BOX 6.2

Egypt and Jordan knit ties around stem cells

More and more developing countries are placing bets on stem cells in their efforts to come up with therapy options for their growing burden of non-communicable diseases. The leading developing countries in this field, China and India, have certainly singled out stem cell research as a promising field to support, but the interest in stem cell R&D in the developing world goes beyond these nations (Lander *et al.,* 2008; McMahon *et al.,* 2010). Middle Eastern countries such as Iran and the Gulf States are prioritising R&D efforts focussed on various kinds of stem cells (El-Awady, 2008; Greenwood *et al.,* 2006). In this spirit, several institutions in Egypt and Jordan have started to collaborate in their promotion of stem cell R&D. For example, the Tissue Engineering Centre in the Faculty of Dentistry (Alexandria, Egypt) and the National Centre for Diabetes, Endocrinology, and Genetics (NCDEG) (Amman, Jordan) started in 2007 to draft a new protocol for the use of stem cells in the treatment of the hepatitis C virus. The focus has been, thus far, on mutual training through courses on advanced techniques for stem cell purification. Joint research is also starting between Ain Shams University (Cairo, Egypt) and the NCDEG on new methods for treating diabetes with stem cell therapy. Recent statistics shed light on why these specific health concerns are the focus for collaboration in stem cell research between Egypt and Jordan. Egypt has the highest prevalence of hepatitis C in the world, and although Jordan has relatively few cases compared to other countries in the region, its rates are still higher than those of developed countries (Mohsen and Norris, 2007; WHO, 2002). Jordan and Egypt are also currently among the 10 countries with the highest incidence of diabetes, and estimates for 2025 suggest that this will remain unchanged (WHO, 2009). These statistics bring to light the influence of shared health concerns on topic selection for these collaborations. Also, Egypt and Jordan both have Islamic populations so their ethical views on stem cell research are likely to be closely aligned. Common views and shared ethical attitudes towards using stem cells, especially embryonic stem cells, create an environment that fosters linkages between countries.

equipment, one researcher we interviewed suggested creating a database of all equipment available at centres and institutions in the Arab world. This would lead to widespread knowledge about existing facilities, and ultimately lead to their better use and more collaboration.

6.4.2 Challenges to the collaboration

Our interviewees emphasised that many challenges limited South-South health biotechnology collaboration or lead to difficulties engaging in it.

1. *Lack of governmental support and direction for health biotechnology research:* Many of the interviewees expressed concerns that health biotechnology was not well supported, either in Egypt or Jordan, and that this limited the extent to which these countries could leverage the potential of South-South collaboration. As a Jordanian researcher stated, 'The scientists in our countries are good but without the proper funding and the equipments and the support they will not produce anything...Science needs to gain more respect in the region.' More investment in this field is needed in order for researchers to be able to carry out their work. A special challenge for collaboration was lack of dedicated time for research. Many of the interviewees were university professors who said that teaching demands interfered with their research and research collaborations. They also expressed concerns that their governments did not set clear priorities for health research. Had these existed, they would have the effect of channelling South-South collaboration and increasing its impact. In our research, we found that although it is a general belief among researchers that collaborations are centred on the health needs of the local populations, there is significant variability in what these priorities are thought to be.

2. *Limited team spirit:* Some of our respondents maintained that a significant limitation to South-South collaboration was the low emphasis put on team work by researchers in the Arabic region. Researchers were more intent on working individually than forming international teams. Even domestic collaboration was in short supply because of a lack of team spirit. This concern was

echoed in earlier research on health biotechnology innovation in Egypt, where domestic collaboration was found to be limited due to a preference for working alone, a lack of trust, and a culture of 'academic secrecy' (Abdelgafar *et al.,* 2004).

3. *Lack of awareness of available expertise:* Interviewees suggested that a significant barrier to optimising the potential for research collaboration internationally is lack of awareness about the expertise available elsewhere within the field. Without at least a general idea of what research is being carried out where, the potential to find suitable collaborators decreases. As an example of a way to overcome this challenge, Jordan's Virtual Centre for Biotechnology, established by the Higher Council for S&T, maintains an active database of researchers and their areas of interest. Furthermore, it helps to arrange brainstorming meetings to bring potential collaborators together, encourage potential collaboration projects, and provide small-scale funding to those projects that show the greatest potential.

6.4.3 Impacts of the collaboration

So far the main impact of Egypt's research collaboration has been capacity building in the health biotechnology field. Considering that Egypt is relatively new to the field, capacity building is an important impact of the collaborations. The country's collaborations with both China and Jordan have, for example, resulted in the development of formalised training opportunities within the three countries. These range from exchange visits and workshops or courses, to undergraduate and graduate-level programmes within university settings. Other capacity-building efforts include more informal knowledge dissemination by taking part in joint research, where partners can share knowledge and learn new methods from each other. Egypt's South-South collaboration has, however, to a limited extent led to co-publications in international peer-reviewed journals. As Egypt seems to be increasing its emphasis somewhat on South-South collaboration in science and technology, it is possible that Egyptian co-authored papers in international peer-reviewed journals will increase in the not-too-distant future. Still, one interviewee pointed out that South-South collaboration did not necessarily lead to co-publications, since the involvement of Northern authors was more likely to be needed for publications' acceptance in international journals.

So far South-South research collaboration in Egypt has not had impacts in terms of contributing to the development of new products/services in the health biotechnology field. We have not been able to identify a case of Egypt's South-South collaboration involving universities or public research groups in the health biotechnology field, which have generated knowledge that has been developed further in preparation of new health products and services. Considering that Egyptian researchers have only recently become involved in health biotechnology and even more recently started to engage to a limited extent in South-South collaboration, it is unrealistic to expect the collaborations to have had much impact on innovation in this field. Instead, Egypt's entrepreneurial collaboration involving firms has had a stronger impact on the preparation of health products, as we will discuss below.

6.5 Collaborations in entrepreneurial activities

Egypt does not have dedicated biotechnology firms, but its pharmaceutical industry is slowly breaking into the biotechnology sector and manufacturing biotechnology products. While Egypt's research collaborations in health biotechnology are strongly focussed on developed countries, it was the only developing country surveyed that was more active in South-South than in South-North entrepreneurial collaboration (Melon *et al.,* 2009; Thorsteinsdóttir *et al.,* 2010). As we have discussed above, its main Southern partners for entrepreneurial collaborations are China and India, but it also has some collaboration with neighbouring countries such as Jordan. The best example of Egypt's South-South collaboration is the collaboration between an Egyptian company and a Chinese firm to enable the production of recombinant insulin in Egypt. Recombinant insulin was previously imported and, as a result, was often in short supply in the Middle Eastern country. The partnership involved the transfer of technology to produce recombinant insulin from the Chinese company Dongbao (Shanghai, China) to the Holding Company for Biological Products and Vaccines (VACSERA) in Giza, Egypt. As a result, Egypt now has a facility that can produce recombinant insulin locally, and diabetics in the country have a reliable and readily accessible supply of insulin that is cheaper than the imported product. The technology transfer from China has thus considerably benefitted the Egyptian health system. As economic and political turmoil can lead to an unsteady supply of important health products, self-sufficiency is far from being a trivial goal for developing countries.

6.5.1 Reasons for the collaboration

1. *Market access.* The results of both our survey and case-study research point to 'access to markets' as a common motivation behind Egypt's South-South entrepreneurial collaborations. Firms collaborate in marketing and distribution in order to expand their markets. From looking at the survey, it is clear that in most cases, Egyptian firms are gaining access to export markets, rather than the reverse. Egypt's main export countries are predominantly also its neighbours (e.g. Jordan). Still, the domestic Egyptian market for health products is large and expanding (ADE/DOL, 2004). There seems, therefore, to be an increased interest in biotechnology firms by countries such as China in gaining access to the Egyptian market. As a representative from a Chinese biotechnology firm put it, 'We go to Egypt as part of our strategic plan to expand into the developing world. We are willing to collaborate with Egypt as it is easier than the markets of the USA and Europe.' Egypt may be an easier market to penetrate due to lower levels of competition from local companies, as many of these are still just beginning to enter the marketplace of biotechnology products, or because of the over-the-counter availability of many products that might elsewhere require stricter distribution procedures.

2. *Gain knowledge:* Our case-study research showed that gaining access to knowledge is an important reason for the Egyptian firms' South-South health biotechnology collaboration. Our survey found that firms in China and India provide Egyptian pharmaceutical firms with biotechnology knowledge. Egyptian generics firms are entering into the manufacturing of biotechnology products through technology transfer arrangements. Some interviewees said that technology transfer was preferable from China rather than developed countries because the cost was lower. The Nile

Company for Pharmaceutical and Chemical Industries from Egypt collaborated with a Chinese group on the technology transfer to produce interferon within Egypt. The transfer of the technology was successful and Nile was able to scale up production and continues to manufacture the final product independent of further involvement from their partners in China. As we can see in Box 6.3 there is scope for Egypt to learn from China in diverse activities related to the health biotechnology field, including lessons on how to promote health biotechnology development through science park development.

BOX 6.3

Egypt and China collaborate in science park development

The Mubarak City for Science and Technology (MuCSAT) (New Borg El Arab City, Alexandria, Egypt) is the product of an agreement signed in 1996 between the Ministry of State for Scientific Research of Egypt and the State Science and Technology Commission of China (SSTCC). It was during a visit by the Vice-Minister of SSTCC, Ms Deng Nan, that the foundation for a future science park was laid and cemented with the promise of around a US $0.5 million donation from China. Among other key areas of science and technology, the park has a heavy emphasis on biotechnology development and commercialisation. In fact, upon its inauguration in August of 2000, one of the first research centres to be established was the Genetic Engineering and Biotechnology Research Institute (GEBRI).

The park merges its focus on biotechnologies with the objective of cooperating with both national and international technology institutes, in its continued collaboration with China's Shenzhen High-Tech Industrial Park (SHIP). SHIP also has a strong component of promoting China's technology advancement through international technical development. In 2003, Shenzhen's Department of Science and Technology provided financial support to MuCSAT in support of new initiatives. South-South collaboration in science park development is an attempt to learn from each other about strategies to commercialise health biotechnology.

3. *Meet local need for affordable health products:* Egypt's aspiration to lessen its dependency on importing important health products

has been a driver for its South-South collaboration in health biotechnology. The need to rely on outside sources for health products in high demand can leave developing countries in a vulnerable position, especially under conditions of economic or political instability. As mentioned above in Egypt, human insulin has at times been in short supply, which has made it difficult for its diabetic population. To help overcome this, VACSERA in Egypt acted to produce recombinant insulin locally, in collaboration with China. Summarising the importance of the collaboration in this particular instance, an Egyptian interviewee stated, 'When Egypt faced a problem with imported insulin from the developed countries, there was only the door of China open to overcome the insulin deficiency crisis.' Now, Egypt is capable of meeting its own demands and ensuring a more affordable product is available to its population in need.

6.5.2 Challenges for the collaboration

1. *Immature regulatory system:* The main challenges for Egypt's South-South health biotechnology collaborations are associated with the country's own drug regulation system not working efficiently. It is still relatively young and its inefficiency can interfere with the success of South-South collaborations. As an Egyptian entrepreneur said: 'There are always problems with the regulatory agencies when registering new drugs. So developing a scientific file is always troublesome.' Interviewees stated that regulatory procedures often take a long time, and the process is tedious to the point of creating tension and distrust among collaborators. Furthermore, the regulatory system is better equipped to deal with more traditional, small molecule drugs rather than biotechnology drugs, and needs greater capacity in order to accommodate the latter. Until recently, biotechnology products needed to be sent outside the country for quality testing, adding both time and expense to the process. Now, however, there is a new division of the Egyptian National Regulatory Authority (ENRA), which is in charge of biotechnology products and applies WHO and other

international guidelines. This may, in time, help alleviate some of the regulation-related delays.

2. *Lack of trust among firms:* Some interviewees felt that lack of trust between firms limited their collaborations. Collaborators do not freely share their specialised knowledge due to fears that they will be taken advantage of. The reluctance to share knowledge was partly attributed to an immature intellectual property right system. As one interviewee notes,

   > Innovation in developing countries needs to be encouraged and supported by strong intellectual property rights systems that are beneficial to researchers, industry, and the country. A poor intellectual property rights system resulted in absence of trust between researchers and the industry.

3. *Difficulties in communications and cultural differences:* Whereas collaboration between Egypt and Jordan is facilitated by a similar culture and a shared language, differences in these areas can make the collaboration between Egypt and China more challenging. There was a consensus between interviewees in Egypt and China that the language barrier and dissimilar cultures posed difficulties for their collaboration, and led to miscommunications and a lack of trust. Despite of these difficulties the Egyptians and the Chinese have been able to overcome these challenges and engage in fruitful collaborations discussed above. These countries, therefore, have had to put special efforts in communication and understanding each other's cultures to foster mutually fruitful collaborations.

6.5.3 Impacts of the collaboration

In line with the emphasis on gaining market access by Egypt's South-South entrepreneurial collaborations in health biotechnology, increased exports were an important outcome of these collaborations. By working together, the Egyptian firms gain access to, for example, Jordanian distribution networks. Joint products on the market were thus the main outputs from Egypt's collaboration that our survey on South-South entrepreneurial collaboration revealed. As developing countries' health biotechnology products are typically less costly than those produced by developed countries, South-South health biotechnology collaboration may lead to more affordable products reaching Southern markets.

As we observed that technology transfer was an important aspect of Egypt-China collaboration, capacity building in this field has been a significant outcome of the collaboration. Based on interview evidence, we argued above that increased capacity has strengthened the supply of important health products, such as recombinant insulin, in the Egyptian market, and lessened the country's reliance on often more costly imports. For the Chinese firms, the main impact has been revenues from licensing their technologies to Egyptian firms. Still, our examination of Egypt's South-South collaboration in health biotechnology has not been able to identify cases of collaboration leading to new-to-the-world innovation in this field. This is further supported by our survey results which show that none of the collaborations have led to new products in the pipeline or joint patenting. This is not altogether surprising as Egypt has a weak innovation record in the health biotechnology field. As a result, we conclude that apart from capacity building, the innovation impacts of Egypt's South-South collaboration have been modest.

6.6 Conclusions

Our case-study research on Egypt's health biotechnology collaborations with Jordan and China has identified several important findings that shed light on the potential and limitations for South-South collaboration in this field. The main messages from this research are:

South-South collaboration in Egypt's health biotechnology sector is likely to increase significantly within the next few years. Egypt is a relative newcomer to South-South health biotechnology collaboration, and it is only in the new millennium that Egypt has been signing science and technology agreements with other developing countries, and allocating resources towards South-South collaboration. Some of these agreements include health and biotechnology foci, and there is thus reason to believe that South-South collaboration in this sector will start to take off with increased opportunities for joint R&D.

Developing countries need to single out health biotechnology and support its development in order to be able to leverage South-South collaborations. Our research has shown that there is, at times, tepid interest in South-South health biotechnology collaboration by Egyptian researchers; in their international collaborations they are almost solely focussed on developed countries. Egyptian researchers seek access to the advanced technologies and richer resources of developed countries to fund their joint research projects. Developing countries have to invest in research field and prioritise areas of importance in order to maximise the potential of South-South collaboration. Further, the collaboration would not have much impact unless it is tightly integrated into wider innovation plans of the participating countries.

Shared health problems and smaller technological gaps facilitate South-South collaboration. A strong motivation for South-South health collaboration is the fact that many developing countries share the same health problems, often different from those encountered by developed countries. Furthermore, as the technological expertise of developing countries is often more like each other's than the developed countries, the solutions developed through South-South collaboration are likely to be more cost-effective and better adjusted to the environments of developing countries.

Technology transfer between developing countries can lead to significant capacity-building benefits to the health system. As we have observed in the cases of Egypt-China entrepreneurial collaboration, the technology transfer has not only had an impact of increased value added to production in Egypt, but it has also led to a better supply of essential health products for local populations. As economic and political turmoil can lead to an unsteady supply of important health products, self-sufficiency is an important goal for developing countries and can potentially be realised through South-South collaborations. Countries can build networks to ensure that their allies have access to products that keep their populations healthy and protect against some of the detrimental health effects of economic recessions and political turmoil.

Our case study on Egypt's South-South collaboration has shown that ties with developing countries can contribute to capacity building in the health biotechnology field. This capacity building can be focussed both on research leading to increased knowledge of Southern countries' health problems and also, according to our interviewees, on entrepreneurial collaboration leading to greater availability of affordable health products. Still, learning is strengthened and made more effective by government support to the health biotechnology sector and prioritisation of important health problems that can channel the South-South collaboration. Only by making local investment in this field can countries leverage their joint collaboration.

References

Abdelgafar, B., H. Thorsteinsdóttir, U. Quach, P.A. Singer and A.S. Daar (2004). "The emergence of Egyptian biotechnology from generics", *Nature Biotechnology* 22: DC25-DC31.

Abedini, J. and N. Peridy (2008). "The greater arab free trade area (GAFTA): An estimation of trade effects", *Journal of Economic Integration* 23(4): 848-872.

ADE/DOL (2004). *Egypt's pharmaceutical sector: Survival and development strategy incorporating results and conclusions of review activity*. Development Options Limited/ ADE and Industrial Modernisation Centre.

Boumedjout, H. (2008). "Algerian research to benefit from Egyptian experience", *SciDev.net* 13. March 2008.

COMESA (2010). *History of COMESA*. The Common Market for Eastern and Southern Africa.

DAAD (2010). *New agreements on research funding and academic mobility as a follow-up to the German-Egyptian year of science and technology 2007: New scholarship programs between the DAAD and the MHESR*. Cairo: Deutscher Akademischer Austauschdienst.

El-Awady, N. (2008). "Gulf states embrace stem cell technologies at home and abroad", *Nature Reports Stem Cells*, 17. January.

ESIS (2006). *Egypt, Mubarak and Africa*. Cairo: Egypt State Information Servie.

Frew, S.E., R. Rezaie, S.M. Sammut, M. Ray, A.S. Daar and P.A. Singer (2007). "India's health biotech sector at a crossroads", *Nature Biotechnology* 25(4): 403-417.

Frew, S.E., S.M. Sammut, A.F. Shore, J.K. Ramjist, S. Al-Bader, R. Rezaie, A.S. Daar and P.A. Singer (2008). "Chinese health biotech and the billion-patient market", *Nature Biotechnology* 26(1): 37-53.

Greenwood, H.L., H. Thorsteinsdóttir, G. Perry, J. Renihan, P.A. Singer and A.S. Daar (2006). "Regenerative medicine: New opportunities for developing countries", *International Journal of Biotechnology* 8(1-2): 60-77.

IMC (2008). *Top industrial companies in Egypt (by sector)*, Industrial Modernization Center. Available at *www.imc-egypt.org/en/imc/index.asp*.

INSouth (2009). *Egypt*: INSouth (an Intellectual Network for the South).

Khader, J.S. (2008). *Twinning program between KHCC-Amman and NCI-Cairo: A unique model in the Arab world*.

Lander, B., H. Thorsteinsdóttir, P.A. Singer and A.S. Daar (2008). "Harnessing stem cells for health needs in India", *Cell Stem Cell* 3(1): 11-15.

McMahon, D.S., H. Thorsteinsdóttir, P.A. Singer and A.S. Daar (2010). "Cultivating regenerative medicine innovation in China", *Regenerative Medicine* 5(1): 35-44.

Melon, C.C., M. Ray, S. Chakkalackal, M. Li, J.E. Cooper, J. Chadder, W. Ke, L. Li, M.A. Madkour and S. Aly (2009). "A survey of South-North health biotech collaboration", *Nature Biotechnology* 27(3): 229-232.

MFA (2008). *A report on Egyptian-Chinese bilateral cooperation.* Cairo, Egypt: Ministry of Foreign Affairs. Available at *www.mfa.gov.eg/MFA_Portal/.*

MHESR (2010). *German Egyptian research long-term scholarship*. Egypt: Ministry of Higher Education and Scientific Research.

Mohsen, A. and S. Norris (2007). "Hepatitis C (chronic)", *Clinical Evidence*, October.

Quach, U., H. Thorsteinsdóttir, J. Renihan, A. Bhatt, Z. Aesch, P. Singer and A. Daar (2006). "Biotechnology patenting takes off in developing countries", *International Journal of Biotechnology* 8(1): 43-59.

Sawahel, W. (2006). "Egypt and Algeria sign science deals with China", *SciDev.Net* 9. November.

———. (2007a). "Egypt and Kazakhstan join forces for research", *SciDev.Net* 16. March.

———. (2007b). "Egypt and Sudan strengthen scientific ties", *SciDev.Net* 24. April.

———. (2008). "South–South industrial centre opens in Egyp", *SciDev.Net* 3. July.

The World Bank (2010a). *World development indicators database: GDP growth (annual %).*

———. (2010b). *World development indicators database: Research and development expenditure (% of GDP).*

Thorsteinsdóttir, H., A.S. Daar, P.A. Singer and E. Archambault (2006). "Health biotechnology publishing takes-off in developing countries", *International Journal of Biotechnology* 8(1): 23-42.

Thorsteinsdóttir, H., C. C. Melon, M. Ray, S. Chakkalackal, M. Li, J. E. Cooper, J. Chadder, T. W. Saenz, M. C. de Souza Paula and W. Ke (2010). "South-South entrepreneurial collaboration in health biotech", *Nature Biotechnology* 28(5): 407-416.

UNESCO (2009). *UNESCO Institute for Statistics fact sheet: A global perspective on R&D.*

WHO (2002). *Hepatitis C, Fact sheet*. World Health Organization.

———. (2005). *Health-related research institutes on genomics and biotechnology in the WHO Eastern Mediterreanean Region*. Cairo: World Health Organization; Regional Office for the Eastern Mediterranean.

———. (2009). *Introduction of diabetes in the Eastern Mediterranean Region*. World Health Organization.

7 A Growing Southern Agenda: India's South-South Health Biotechnology Collaboration

AUTHORS: *Sachin Chaturvedi and Halla Thorsteinsdóttir*

7.1 Introduction

Since India gained independence in 1947, its foreign policy has emphasised collaboration with fellow developing countries. The charismatic leadership of Jawaharlal Nehru included a special focus on both science and technology and on South-South cooperation. It was under his leadership that the Southern economies met in 1955 and organised the First Afro-Asian Conference at Bandung (Indonesia) discussed in Chapter 1, which can be said to mark the beginning of formal collaboration between developing countries. This meeting convened member countries 'to provide technical assistance in the form of experts: exchange of knowledge and establishment of national and where possible, regional training and research institutions for importing technical knowledge and skills' (MEA, 2005). Despite recognising the importance of South-South collaboration, India has been slow to engage in collaboration. Partly because of limited resources, India's South-South collaborations have largely been confined to a narrow set of activities.

India's economy has recently been building up steam. Its economic growth in the 2000s has been one of the highest in the world, reaching nearly 9 per cent annual GDP growth rate in the last few years (MoF, 2010). India has one of the largest populations of all countries in the world, reaching

1,367.2 billion in 2007 (UNDP, 2010). Its economic growth has resulted in considerable expansion of India's consumer market, and by 2025 the size of the market is expected to reach 500 million (Ablett *et al.,* 2007). India, is an emerging economy which has shown consistent efforts to enhance R&D expenditure (OECD, 2009). As percentage of gross national product (GNP) in 2005–06, R&D expenditure increased from 0.81 in 2002–03 to 0.89 (DST, 2008). Drugs and pharmaceuticals occupy the first place, with a share of 45.1 per cent in the total industrial R&D expenditure by the private sector (DST, 2008). As a result of all these factors, there is fresh impetus for India to pursue South-South cooperation. The growing polarisation within global platforms, such as the World Trade Organization (WTO), has encouraged new, smaller, and more dynamic groupings, such as that of IBSA. Apart from trade, IBSA also actively cooperates in other areas like S&T etc. For instance, they have identified research in biomedical technology as a priority area for further collaboration in the grouping, and have established new fellowships which are exclusively dedicated to these economies. With India's recent economic growth, its opportunities for generosity have increased.

However, at this stage it appears that this strength may have limitations as well. It seems that—at least in the medical biotechnology sector—'South,' for India, tends to refer to the *nouveau riche* class of Southerners. So, there is greater cooperation with fellow 'emerging' economies, but the wider community of the Third World is being left out, even though India once embraced its leadership at fora such as at the Non-Aligned Movement and the UN Group of 77 (G-77). As a key science and technology policy analyst in Delhi remarked: 'With expansion in resources and technological accomplishments, India has to do tight rope walking. We should not allow ourselves to be absorbed in the OECD world nor abandon the developing world.'

The last five Annual Reports of the Department of Biotechnology (DBT) show that each year, the reported number of collaborations it supports with developing countries has declined. In the latest year, 2008–09, DBT reported only a single collaboration, a surprising statistics given that a few years ago the DBT itself had collaborations with several developing countries. The Department of Science and Technology (DST), along with a few national institutions, are currently holding fort in the realm of biotechnology diplomacy. The DST is also providing support to the IBSA process.

With health biotechnology emerging as an important area in its national R&D strategy, India has built capacity in this field since the early 1980s, and it has growing public and private sector strengths in the field (Chaturvedi, 2005 and 2008; Kumar *et al.,* 2004). As Indian entrepreneurs expand their market reach, the Indian Government can use its strengths both to share its capacity with developing countries that are weaker in the field, and by strengthening joint research and innovation efforts with developing countries active in health biotechnology.

BOX 7.1

Case-study research on India's South-South collaborations: Methodology

As described in Chapter 1, the case studies relied on multiple sources of data, including scientometric analysis of co-publications, a survey of firms about their collaborations, and interviews with 50 experts in the chosen countries (Table 7.1). These data are further supplemented through document analysis of policies and background literature, and any other available statistics of relevance to the topic.

In order to understand the level, impact, challenges, and potential of India's health biotechnology collaborations with its Southern partners, we carried out case-study research on both research and entrepreneurial collaborations in the health biotechnology field. We focussed particularly on the partnerships between India and Brazil, another emerging economy, and India and Bangladesh, a country that is just embarking on its biotechnology development. We studied a total of two cases and defined each case as the bilateral collaboration between India and Brazil *versus* India and Bangladesh. This research design gave us an opportunity to study both India's collaboration with a country advanced in the field and also its collaboration with a newcomer in the field. As discussed in Chapter 4, Brazil is one of the leading developing countries in the health biotechnology field and started the development of the sector in the 1970s. In comparison the Government of Bangladesh did not adopt a formal National Biotechnology Policy until in 2006. It includes an emphasis on health biotechnology and the establishment of a National Technical Committee focussing on medical biotechnology. This policy inclusion on biotechnology has, however, not been accompanied by increased allocation of resources towards biotechnology development.

India's political emphasis on South-South collaboration and its recent economic growth have made it important to examine the extent to which India has moved beyond rhetoric and is emphasising South-South collaboration in health biotechnology. Here, we examine, through case-study research (Box 7.1), India's South-South health biotechnology collaborations with a special focus on its collaborations with Brazil and Bangladesh. We identify the driving forces behind India's collaborations, the challenges they have met along the way, and the impacts of the collaborations thus far.

TABLE 7.1

Breakdown of number of interviewees in case-study countries

Countries	Number of interviewees
India	24
Bangladesh	8
Brazil	18
Total	50

7.2 Government interest and support

To explore the policy context of India's South-South collaboration, we look at the emphasis India's government has placed on collaborating with other developing countries. After independence, India began to tailor its foreign policy towards fostering closer ties with developing countries in lieu of providing traditional grants and aid. India has focussed extensively on South-South cooperation through training and scholarships, with a structured programme under the Ministry of External Affairs called the Indian Technical and Economic Cooperation Programme (ITEC) (MEA, 2008). As a part of this programme, Indian experts have travelled to other countries to provide training in various areas such as telecommunications, transportation, medicine, and public health (Chaturvedi, 2008). This cooperation has consisted primarily of first-generation technologies, but in a few countries, it has been extended to high-tech sectors as well. Presently, India allocates around US $10 million a year to ITEC, and has supported development in almost 150 countries (Chaturvedi, 2008).

Under the Ministry of Science and Technology, the DST is also strongly focussed on South-South cooperation. The department is guided largely by efforts to strengthen India's own S&T sector while simultaneously promoting international initiatives that are likely to reflect its S&T leadership in empowering other developing countries (DST, 2009). In total, India currently has 73 bilateral S&T cooperation agreements, including with both Bangladesh and Brazil, and uses a range of instruments in its collaborations, including exploratory scientific missions, workshops, joint research projects and development centres, and advanced training fellowships (DST, 2010). Aside from these bilateral agreements, the DST engages with other countries through a series of multilateral and regional ties. For example, New Delhi is currently home to the Secretariat of the Centre for Science & Technology of the Non-Aligned and Other Developing Countries (NAM S&T). Among its other objectives, it strives to 'promote the fullest possible and mutually beneficial collaboration among scientists and technologists and scientific organizations from non-aligned and other developing countries' (NAM, 2010). In terms of DST, the main drivers of international collaboration have been technology diplomacy, synergy and acquisition. There has been a relatively limited emphasis on joint South-South research projects, and much stronger focus on capacity building.

India also takes part in multilateral initiatives aimed at encouraging South-South collaboration. In the biotechnology field it takes part in the International Centre for Genetic Engineering and Biotechnology (ICGEB), a UN initiative focussed on strengthening research and training in molecular biology and biotechnology for developing countries' needs. India has been a member state since its very beginning, and the first ICGEB centre in a developing country was established in New Delhi in 1988 with co-funding from the Indian government. The centre supports both training and research activities in India and in other developing countries.

We see a relatively strong capacity-building focus in India's multilateral initiatives. For example, it directs 10 per cent of its UNDP support towards fostering South-South activities, and, together with UNESCO, established the UNESCO Regional Centre for Biotechnology Training

& Education in New Delhi (Parthasarathi, 2000; Sawahel, 2005). This centre combines research and training in biotechnology, and focusses primarily on capacity building in South Asia. India has also promoted biotechnology development within the ASEAN forum to promote human resource development and sustainable use of biodiversity and genetic resources. It established the India-ASEAN Institute of Biotechnology (IAIB) in Jakarta, where R&D is carried out predominantly in plant biotechnology, but also in pharmaceuticals and bioinformatics. Additionally, the India-ASEAN collaboration emphasises collaboration in technology management and intellectual property issues in order to help countries pursue product development and commercialisation (Padma, 2005). So we are seeing various governmental activities by the Indian government to support South-South collaboration. However, there is a risk that they are too diverse to have significant impacts. As we observe a lacklustre performance in promoting South-South collaboration even by some ministerial organisations such as DBT, there is reason to believe that more concerted efforts by the Indian government to promote South-South collaboration are needed.

In recent years, India seems to have revitalised interest in South-South collaboration largely though the trilateral, IBSA developmental initiative, which promotes cooperation and exchanges between the three countries. As discussed in Chapter 1, IBSA works in several sectors and is strongly focussed on developing Southern consensus on international issues in order to present a more united front at the international fora, and to promote trade between the three countries. Furthermore, collaboration has begun in the areas of health affairs, and science and technology, with IBSA singling out research cooperation in such fields as tuberculosis, malaria, HIV/AIDs, and biotechnology. The IBSA forum thus serves to focus the countries on joint efforts in health biotechnology research as well as a forum to discuss the elimination on non-tariff trade barriers between the countries. The trade between India and Brazil has expanded rapidly since IBSA started its operation in 2003, from US $589 million to US $2.5 billion in 2006 (Mokoena, 2007). The future will reveal whether the IBSA is successful as a tool for delivering the messages of developing countries, and fortifying their potential to address joint health needs through collaboration.

India's governmental emphasis on collaboration with Brazil is also evident, from the establishment of the so-called Indo-Brazil Science Council (IBSC) in 2007 with the goal of promoting joint R&D projects and other activities which would bring together both research hubs and entrepreneurial centres within the two countries (see Box 7.2). According to interview evidence, both India and Brazil have contributed US $1 million each for joint research projects. The fund is used to fund joint R&D projects, workshops/seminars, and exchange of junior and senior scientists.

BOX 7.2

Emerging Indo-Brazilian relations: Recent key milestones

Brasilia Declaration, June 2003: India, Brazil, and South Africa (IBSA) decide to work together.

Emergence of G-20 at WTO Meeting: India and Brazil work together on issues at WTO, in favour of developing countries.

First Joint Commission Meeting, October 2003, New Delhi: Brazilian Minister of External Relations of Brazil leads Brazilian delegation to India, comprising officials and businessmen. The two governments decide to enhance their collaboration in many areas.

Visit of President Lula da Silva to India, January 2004: President Lula travelled to India with 97 Brazilian businessmen. Meetings with Indian companies took place and a better commercial relationship was established.

India signs Preferential Trade Agreement with MERCOSUR in December 2004.

Governor of São Paulo visits India with a large business delegation in November 2005. Governor met PM and visited the India International Trade Fair, 2005.

Second Joint Commission Meeting held in Brasilia, on February 1 and 2, 2006.

India hosts Third IBSA meeting October 2008 at New Delhi.

With Bangladesh, there is no specific agreement for cooperation in this area but the governmental willingness is reflected through its joint participation in SAARC. It is also expressed through a trilateral cooperation agreement between India, Bangladesh and Nepal led by the WHO to work on *kala azar* or visceral leishmaniasis (VL), the second most deadly parasitic disease in the world following malaria. The disease occurs predominantly in the poor and marginalised communities. During 2000–2002, the reported cases in Bangladesh, India, and Nepal were 24,287, 18,472 and 22,030, respectively. Not all cases may be reported and estimates indicate about 100,000 cases per year in the region (WHO, 2005).

The governments of India and Bangladesh have also undertaken measures to improve entrepreneurial linkages (The World Bank, 2008). The Reserve Bank of India recently liberalised the ban on foreign direct investment (FDI) from Bangladesh which may eventually lead towards a stable and secure trade relation between India and Bangladesh (ANI, 2009). The two countries have signed the new Bilateral Investment Promotion and Protection Agreement (BIPA) that seeks to promote and protect investments from either country in the territory of the other, with the objective to increase bilateral investment flow. With BIPA, Bangladesh and India will henceforth consider each other a most favoured nation (MFN) (Al Abbas, 2009). The total bilateral trade between India and Bangladesh grew from US $2.56 billion in 2006–07 to US $3.616 billion in 2007–08. In order to promote trade further, the India-Bangladesh Chamber of Commerce & Industry was also established in 2007.

In general, it seems that South-South cooperation in science and technology has emerged as a key political constituent for external cooperation of independent India. The focus has been strongly on training programmes, capacity building, and fellowships with very little stress on joint research *per se*. As a result it is not surprising that our research indicates that India's research collaboration with developing countries is relatively small. India's political rhetoric emphasises South-South collaboration and the country shares common challenges with other developing countries that lend themselves to be addressed with joint collaboration. Below, we examine those potentials further by looking at India-Brazil and India-Bangladesh research and entrepreneurial collaborations in health biotechnology.

7.3 Geography of collaboration in India's health biotechnology

7.3.1 Mapping research collaboration

Our mapping of India's research collaboration in health biotechnology shows that Indian researchers collaborate to a limited extent with other developing countries. If we compare the number of South-South co-authored papers India publishes in health biotechnology to the numbers for Brazil and China we can see that Brazil published almost double what India produced or 1,021 papers between 1996–2009 *versus* 604 papers for India. China is also ahead of India in number of co-publications with developing countries in the health biotechnology field with 973 papers for the same period. This finding is consistent with previous research that shows that India is less engaged in international collaboration than other leading developing countries in this field (Thorsteinsdóttir *et al.*, 2006). Still the mapping analysis shows that in recent years India has had a steep increase in South-South co-authored papers, indicating a growing emphasis on collaboration with developing countries (see Chapter 2). It was around 80 papers in 2008 from only around 20 papers in 1996.

The mapping of India's research collaboration shows that China is India's main developing country collaborator in health biotechnology (Figure 7.1). Brazil also has relatively frequent ties to the country, as the second most common country to collaborate with India. It is also notable how much India collaborates with neighbouring Bangladesh, given how new the country is to the biotechnology sector. Bangladesh is India's third most common research collaborator in health biotechnology. At the beginning of the period we studied, India and Bangladesh had almost no co-authored papers, but this increased considerably in the last period studied. From the perspective of Bangladesh, India appears to be an important collaborator. It is its foremost Southern collaborator with 56 co-authored papers for the period studied compared to only 14 co-authored papers with China, its second most common Southern collaborator.

The main subfields of health biotechnology in which India collaborates with other developing countries are in 'Genetics and Heredity' (64 papers) and 'Microbiology' (34 papers). The relatively strong emphasis on genetics may reflect research collaboration on common genetic lineage or the effect

FIGURE 7.1

The main developing countries India collaborates with in South-South research collaboration in health biotechnology based on the number of co-authored papers between 1996–2009

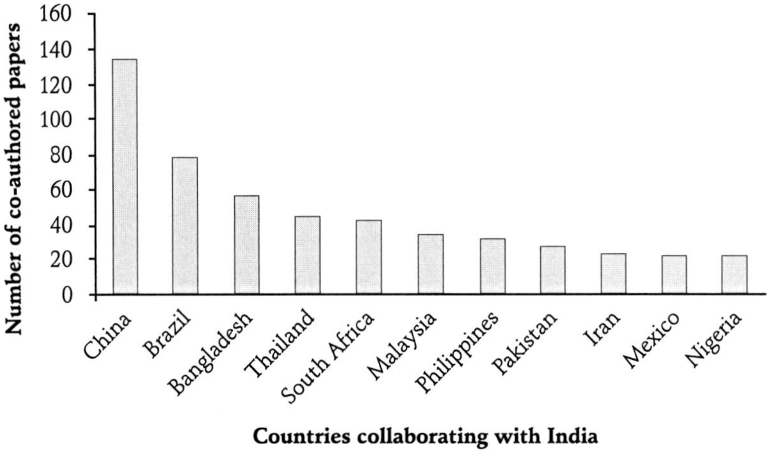

Source: Authors' presentation of data compiled by Science-Metrix using Scopus.

of international forces stimulating international collaboration in this field, for example, through the Human Genome Project and related initiatives.

India has around 4,000 co-authored papers for the period between 1996 and 2009 with high-income countries. These have increased from around 180 co-authored South-North papers in 1996 to just fewer than 500 such papers in 2008. In general India has had modest international collaboration in the health biotechnology field compared to other emerging economies but it has increased in recent years. China, for instance, has almost 18,000 papers for the same period with Northern authors and Brazil has almost 5,200 such papers. The United States (2,084 papers), United Kingdom (518 papers), Germany (457 papers), and Japan (447 papers) are India's main Northern collaborators.

7.3.2 Mapping entrepreneurial collaboration

The total number of entrepreneurial collaborations India had with other developing countries was 54, which was the third highest number of South-South collaborations of the countries examined in this study.

In comparison Brazil and South Africa had over 60 South-South entrepreneurial collaborations each. Looking at the countries that have at least three entrepreneurial collaborations with India, we see that again, the emerging economies are India's main partners, with China and South Africa in the primary places (Figure 7.2). Brazil and Egypt share the next two places as India's most frequent entrepreneurial partners.

Further survey questions revealed that almost all of India's entrepreneurial activities are based around marketing and distribution activities (73 per cent) and almost 21 per cent focus on manufacturing and R&D. India's South-South collaboration in this field reflects a strong commercialisation focus that we will examine further in a later section. China is both India's main research and entrepreneurial collaborator and that is why we researched China-India collaboration in Chapter 5 on China's South-South collaboration in this book. Both the countries that we focus on in this chapter, Brazil and Bangladesh, feature prominently as India's health biotechnology collaborators and it is interesting to assess and analyse the reasons, challenges, and impacts of these collaborations.

FIGURE 7.2

The main developing countries India collaborates with in South-South entrepreneurial collaborations in health biotech, based on a survey of firms

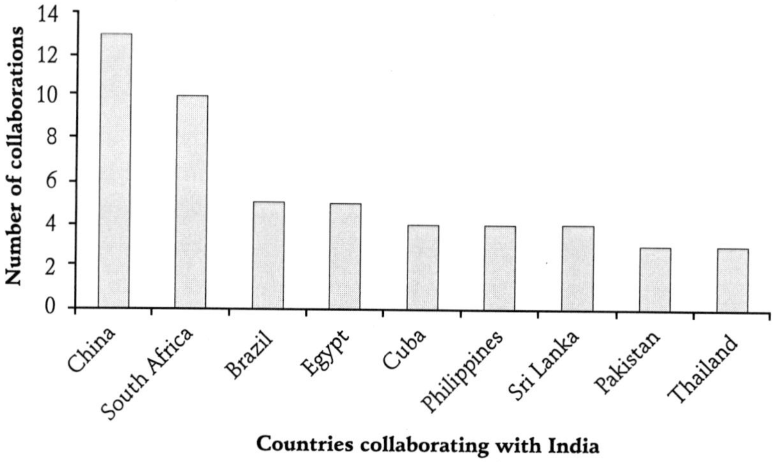

Source: Authors' presentation of data compiled by own survey.

7.4 Collaboration in research activities

Indian researchers have had modest collaboration in health biotechnology with researchers in other developing countries but their South-South collaborations have been increasing in the last few years. In general India's health biotechnology collaborations are strongest in the 'Genetics' and 'Microbiology' subfields of health biotechnology. Some examples of India's South-South collaboration are the collaboration between Jawaharlal Nehru University (New Delhi, India) and University of São Paulo (São Paulo, Brazil) on the genetics of leishmaniasas; the Bose Institute (Kolkata, India) and Fiocruz (Rio de Janeiro, Brazil) work together on researching the genetic composition of different cholera strains. Cholera is also an active topic of India-Bangladesh collaboration. The National Institute of Cholera and Enteric Diseases (Kolkata, India) and the International Centre for Diarrhoeal Disease Research (Dhaka, Bangladesh) together research cholera and other diarrhoeal diseases (see Box 7.3). There are diverse reasons that drive India's collaboration with other developing countries and below we discuss the key ones identified by our case study research.

7.4.1 Reasons for the collaboration

Based on the responses from our interviewees, it appears that India's South-South collaborations are centred around:

1. *Common interests stemming from shared health concerns:* Our interviews revealed that the most pervasive driver of India's South-South collaborations is the desire to work with other developing countries towards finding realistic solutions for common health threats. A good example of a common health threat between India and Brazil is HIV/AIDS. Most of the research done on the HIV virus in the world has been carried out on subtype B. But in India and Brazil, there is a heavier prevalence of subtype C virus and, as a result, an impetus for India and Brazil to research that subtype together. Shared health concerns are reflected in the prioritisation of India's governmental collaboration programmes with Brazil, where the focus seems to be heavily on communicable diseases. Under the IBSC, the foreign ministers of India and Brazil, at their

joint meeting in 2007 in New Delhi, set forth a prioritisation list. Out of some 200 proposals, received in response to a call, nearly 20 projects are approved and the majority are in the health technology sector. Both sides have identified malaria, leishmaniasis, HIV, tuberculosis, and leprosy as target diseases for joint research. Since leishmaniasis is a shared health problem between India and Brazil, several groups in the two countries have focussed on researching this health issue and they have joint publications on this topic. The strains of leishmaniasis are, however, quite different in the two countries, which limits further potential for collaboration. The work in the area of HIV is in a preliminary stage of joint research. These projects are being coordinated by the two governments; from the Indian side it is the Department of Science and Technology, and from Brazil it is the CNPq.

BOX 7.3

South-Asian neighbours partnering to combat regional enteric diseases

The National Institute of Cholera and Enteric Diseases (Kolkata, India) and the International Centre for Diarrhoeal Disease Research (Dhaka, Bangladesh) are internationally recognised centres of excellence in the field of diarrhoeal diseases. Diarrhoeal diseases are the second most common cause of death among young children in developing countries. The two institutions collaborate closely with each other and work with international partners in studying molecular pathogenesis of enteric disease agents such as Escherichia coli, Vibrio cholerae, Shigella dysenteriae, and rotavirus strains. In addition to collaborating on world class academic research on tropical pathogens, they are also involved in developing locally relevant health technologies to serve their populations. Helicobacter pylori is a gastric pathogen implicated in gastric cancer and is prevalent in South Asian populations. In 2004, collaborating researchers from the Indian and Bangladeshi institutions, along with US scientists, developed a simple but novel multiplex PCR assay for rapid detection of Helicobacter pylori infection and virulence genes. As the assay does not require the culturing of strains, the pathogen genotypes can be obtained directly from gastric biopsy specimens. This can save time as well as expensive reagents and instrumentation which is a feature of particular value for laboratories in developing countries.

India's collaborations with Bangladesh have also focussed predominantly on tackling infectious diseases of particular concern to the individual countries (Box 7.3). Cholera is a significant health problem in Bangladesh and Eastern India and, as a result, the two countries have collaborated on health biotechnology research on this health problem, such as the genome of cholera. In fact this is one health concern in which researchers from Bangladesh, India and Brazil have developed informal research networks. As Bangladesh happens to have an international organisation for advanced research, The International Centre for Diarrhoeal Disease Research, Bangladesh (ICDDR,B, Dhaka) and the country is generally new to biotechnology research, it is not altogether surprising that the Indian collaboration almost solely involves experts from this centre instead of domestic institutes.

2. *Complementary expertise:* Our interviewees for this study stressed that complementary expertise was an important reason for their collaboration. As summarised by an Indian researcher, 'The major reason to collaborate is complementing expertise and commonality in scientific thrust.' Together, scientists from both India and Brazil were said to be strengthening their ability to carry through with projects that otherwise might have stalled due to a lack of knowledge. In an example from the work between India and Brazil, one Indian scientist explains, 'I was looking for a thermodynamic explanation, [while] the collaborator did work on hydrocarbons and [was] looking for applications, which I provided. He had a solution in search of an application, and I had an application in search of a solution, and we just matched [them].' Our research did not necessarily indicate that India searched for specific health biotechnology knowledge, neither to Brazil nor Bangladesh nor vice-versa, but rather allowed each partner to draw upon their own specific strengths and experiences to contribute towards a common goal.

3. *Access to samples:* Another driver for collaboration is access to clinical samples or strains. This has led to the formation of several collaborations between the two countries, focussed on the need to access these samples for research purposes. Regarding the

collaboration between the Bose Institute (Kolkata) and Fiocruz (Rio de Janeiro), a Brazilian interviewee explains:

> Since there was no repository kept for strains or isolates, there was no way work related to identification of strains could be initiated in Brazil. Meanwhile, scientists working in India on Cholera were exploring gene sequencing of various strains... This led me to contact them for partnership.

India has access to some genetic strains unavailable elsewhere, and as such, it has become an attractive partner for other countries seeking to gain access to these. Brazil is also providing samples to Indian investigators in malaria research. The Amazon people of Brazil have resistance to malaria which is not found in India. Indian researchers are interested in knowing what strains in the body are different and if these can be linked to the causes of the disease. Access to samples of both the parasites, as well as of the human population, to know the factors that are responsible for resistance is important for research. Malaria is yet another example of a shared health problem in India and Brazil that encourage collaboration.

4. *International meetings and publications leading to joint research projects:* During our study, we came across several instances in which international meetings, publications, and such other international opportunities for interaction played a key role in getting various researchers together to eventually launch joint research studies. It was noteworthy to find that a US university based initiative could get Indian and Brazilian scientists together to work on leishmaniasis. Later these two scientists continued to work together through their respective institutions and even promoted further exchange of researchers in the subsequent years and much detailed collaboration on genomics studies on leishmaniasis.

7.4.2 Challenges for the collaborations

A challenge for India's health biotechnology collaboration with Brazil and Bangladesh is the perception that collaboration with the North is much more valuable than South-South collaboration. In the health biotechnology field we would expect a heavier emphasis on collaboration with developed countries as there is more research taking place in the North than the South (see Chapter 2), and the North possesses more of the needed resources to conduct research in this field. Still some of the comments seem to suggest an exaggerated emphasis on North-South collaboration. One Indian researcher stated, for example: 'All Indians have Western fixation and so do the institutes made of them.' This sentiment also seems to be echoed by Brazilian researchers. In speaking with some of our interviewees, we found out that many of the South-South collaborations were spin-offs from previous studies led by the North. In other cases, a Northern party was involved in bringing the two Southern groups together on paper only, and the two groups from developing countries did not even have the chance to communicate with each other. Funding was also more accessible for collaboration with the North than the South. Joint research at times started between India and Brazil but could not continue due to lack of dedicated funding. For example, research on cholera between the Bose Institute (Kolkata, India) and Fiocruz (Rio de Janeiro, Brazil) was confined to couple of joint papers due to lack of resources.

Another challenge identified by our study was that new governmental initiatives to promote South-South collaboration did not seem to align well with existing South-South collaborations. An Indian researcher explained that new initiatives geared towards strengthening South-South collaborations did not favour researchers who have a track record in carrying out collaborations with developing countries but rather more established researchers. Those established researchers generally have extensive research funding from developed countries and collaboration with Southern countries are in no way their priority. This limits the potential of building upon existing experiences of South-South collaboration and furthers fund-driven collaboration at the expense of more genuine and long-lasting collaboration.

A further challenge pointed out by our interviewees was lack of post-doctoral fellowships. There are, for example, only a few fellowships planned as a part of the Indo-Brazil joint research programmes. Governments need to pay more attention to include fellowships, especially at the junior research level, as a part of their support programmes. Related to lack of fellowships was limited formation of research networks that connected the countries we studied. The researchers in these countries do not generally seem to form an informal or formal network, or establish a trustworthy and comfortable relationship with one another. There are various reasons for this such as language challenges and lack of direct airline flights, which make it difficult for researchers to travel to their collaborators' sites.

7.4.3 Impacts of the research collaborations

Gains in knowledge and access to research material are some of the most common impacts or outputs of research collaborations, and they are there for the Indian South-South linkages that we examined, as well. However, as is apparent, from our case study of India, we also see cases where a Southern collaboration has resulted in the development of a health technology product. For example, a diagnostic tool for leishmaniasis identification was created by the Federal University of Paraná (Curitiba, Brazil) in 2003, and later transferred to the Brazilian Centre for Research in Immunological Products (Box 7.4).

Equally as important to getting these products on the market, is ensuring that they remain cost-effective and affordable for the public. Given that financial resources are often more limited in developing countries, for both governments and individual citizens alike, there is more of a pressure in these environments to create health products that are more cost-effective than others already in existence, or to introduce affordable products that are not yet available. Much of the work in India-Bangladesh collaborations is focussed around the potential of creating cholera vaccines that would cost a fraction of the currently available options. Even if a moderately effective vaccine were to become available, it is thought that it would still reduce levels of disease burden substantially, and would be perceived as a great success. By using research collaboration between the two countries, there is the potential—and desire—to make these technologies available to the Indian public in a way that would increase accessibility through affordability.

Through these few cases of product development, and the broader impact of increasing knowledge and capacity through training programmes and researcher exchanges, South-South collaboration is beginning to have a positive effect on the S&T landscape of India and its collaborators.

7.5 Collaboration in entrepreneurial activities

As Bangladesh is only taking its first steps in promoting biotechnology development, it is not surprising that we only found two India-Bangladesh firm collaborations. Bharat Biotech International Ltd. (Hyderabad, India) and Biological E. Ltd. (Hyderabad, India) both have had entrepreneurial collaborations with the ICDDR,B. The Biological E. is collaborating with the ICDDR,B in developing and manufacturing cholera vaccine. It is also collaborating with Austrian firm Intercell AG (Vienna, Austria) which is in Phase II trials of a paediatric vaccine against Japanese encephalitis. Biological E. will manufacture Intercell's Japanese encephalitis vaccine for the Asian markets and will exclusively market and distribute the product in India, Nepal, Bhutan and Bangladesh. Since India's health biotechnology collaboration with Bangladesh is so limited, in this section we focus primarily on India's collaboration with Brazil.

Our survey identified 22 entrepreneurial collaborations in the pharmaceutical sector between India and Brazil. The entry of Indian firms may easily be divided into two distinct phases. The first phase is from 1994–1999, when few firms entered, without much preparation, and they had to face several challenges. According to an interviewee, 'it was not until 1994 that linkages between Brazil and India began to take off. Indian companies neglected Latin America completely and when they started in 1994 with Dr. Reddy's Laboratory (Hyderabad, India), the strategy was not well suited to the Latin American markets.' Initially, linkages were centred heavily on the importation of Indian health products into the Brazilian market. However, with specific encouragement from the Brazilian Government in 1999, the firms were encouraged to take a long view and invest in Brazil. This had an impact. Ties gradually became deeper, and by 2002 the companies started investing in local production in a major way. In this phase, almost all the major Indian companies like Ranbaxy Laboratories (Gurgaon, India), Strides Acrolab (Bangalore, India), Dr. Reddy's Laboratories (Hyderabad, India), Cadila Healthcare (Ahmedabad, India), Wockhardt Ltd. (Mumbai, India), Orchid Chemicals & Pharmaceuticals (Chennai, India), Torrent Pharmaceuticals Ltd. (Ahmedabad, India), Glenmark Pharmaceuticals (Mumbai, India), Unichem Laboratories (Mumbai, India), etc., have already established subsidiaries for manufacturing or marketing in Brazil. Their entry strategies ranged from manufacturing plants, to joint venture alliances, to acquisitions and mergers. For example, the Indian company Glenmark acquired the Brazilian firm Laboratories Klinger (São Paulo, Brazil) in 2004, and set up a subsidiary in Brazil to capture that country's generics markets. Most of the Indian companies have their subsidiaries in Brazil. One of them, Cellofarm (Rio de Janeiro), is among the fastest growing pharma companies in the generics space in Brazil. With two production units and a business of about US $98 million, the company has a growth of almost over 30 per cent annually.

The budding India-Brazil linkages cover various high-tech areas. Brazil, for example, has emerged as a major centre for organ transplants that require immune suppressants, which Indian companies like Biocon (Bangalore, India) have readily supplied. Based on our interviewee data, the future collaborations in high-tech areas are likely to be in areas such as oncology, particularly blood cancer. Intas Biopharmaceuticals (Ahmedabad, India) is an example, of a firm eyeing the opportunities of India-Brazil collaboration in oncology through its Brazilian subsidiary.

7.5.1 Reasons for the entrepreneurial collaborations

Our case-study research showed that the India-Brazil linkages have been growing because of various reasons that include:

1. *Access to markets:* Tapping into the Brazilian and Latin American markets has been the main focus of Indian firms in this sector. This was supported by our survey which showed consensus between the Indian and Brazilian firms that access to the Brazilian markets was a reason for their collaborations. Aiding India's success in this has been its ability to provide high quality drugs and intermediates at very cost-effective prices. Our interviewee indicated that future phases of collaboration would be likely to include more R&D ties, but for now, the country seems highly focussed on the importing and marketing as activities of the collaboration. It was also clear that the Indian firms set up subsidiaries or ties with Brazilian firms not just to access the Brazilian market but, more generally, Latin American markets.

2. *Governmental push:* A major push to increase India-Brazil biopharmaceutical collaborations came in 1997 when the Brazilian Health Minister, José Serra, invited Indian companies to invest in Brazil and use Brazil as a production hub for pharmaceuticals, rather than a mere export destination. This was an attempt to increase Brazil's local pharmaceutical sector. Later on in 2007, Brazil increased import duties on pharmaceutical products that made it difficult for Indian firms to rely solely on exporting their products to Brazil and pushed them rather to set up a Brazilian operation or collaboration (Matthew, 2007). The total Indian investment in Brazil has multiplied in recent years, and has expanded to US $470 million between 1996 and 2006. A large portion of this investment has been made by Indian pharmaceutical companies. As described by one Indian entrepreneur: 'One of the catalytic factor[s] was the promulgation of rules for generics, which were intended to de-bureaucratise the process as much as possible.' Several leading generic companies responded and, with time, the number of companies present has multiplied to almost 22.

3. *The role of expatriates:* Several Indian companies have been recruiting the Indian scientific diaspora from developed countries. Many of these individuals have experience working in firms in Northern countries that have had close ties with emerging markets, such as multinational pharmaceutical firms, and are more keen to establish ties with other emerging countries as a part of their work with Indian firms. Indian companies, such as Dr. Reddy's Laboratories, Wockhardt, and others, are luring these Indian scientists in to head their R&D units. Interviewees suggest that this has brought a completely different approach to the entrepreneurial canvas. The newly recruited Indian diaspora has previous experience in collaboration with Brazil which they gained when they were working in Western countries. Firms are now more open to the unconventional markets of Latin America, and the experience of the Indian scientific diaspora can therefore be put to good use in establishing closer India-Brazil ties.

7.5.2 Challenges of the entrepreneurial collaborations

Our interviewees indicated that changes in the Brazilian regulatory system have posed challenges to India-Brazil collaboration. ANVISA, the Brazilian regulatory agency adopted new bio-equivalence standards in 2003. The new regulation made it mandatory that contract research organisations undertaking bio-equivalence studies need to be approved and certified by ANVISA. These were applied across the board, including to products that had already been approved by the regulatory body. The new set of standards did not stop there—they also shortened the period of validity for registered products, and increased the registration fees for medical and pharmaceutical product imports. These changes have increased the financial and time burden experienced by Indian companies, looking to expand into the Brazilian market.

Delays in patenting in Brazil have also posed a challenge for India-Brazil collaborations. These delays were compounded in 2003 when a new requirement was made for all applications to be sent to, and approved by the Brazilian Patent Office (INPI) and ANVISA. The nature of the situation has caused great frustration among Indian entrepreneurs who are engaged in collaborations with Brazilian groups, and it could be serving to dissuade the possibility of future collaborations with the country.

Though when compulsory licence was instituted by President Lula da Silva in May 2007 for Merck and Co.'s (Whitehouse Station, USA) efavirenz (AIDS drug), the Indian firm Aurobindo (Hyderabad, India) was selected as the key supplier for the active pharmaceutical ingredients (APIs) but still several firms, whom we interviewed, expressed concerns about the growing urge, especially among public sector agencies, to source drugs from local generic companies. As one of them said, 'Brazilian public sector has moved towards national restrictions for purchasing as is very clear from a growing discrimination from API [active pharmaceutical ingredient] purchases of Far-Manguinhos.' This is important in light of the large share of government's purchases in the overall drug market. In our discussion with Far-Manguinhos (Rio de Janeiro, Brazil), it came out that some Asian API consignments were not of the quality expected, hence the change in policy.

A further challenge for the India-Brazil collaboration in health biotechnology are the difficulties in establishing collaboration between Indian firms and the public research institutions in Brazil. As the latter are the main performers of biotechnology research in the country, the lack of linkages has quenched the innovation potential of the collaborations. Indian companies have not been aware of the expertise within these centres, and vice-versa.

BOX 7.4

Effective diagnosis on the cheap

Brazil has a reputation of having strong public sector research in health biotechnology, which makes important contributions to knowledge production but which is typically not harnessed for development of health products or services. Linkages and knowledge flow between its universities and industries have been tenuous, which hinders its innovation. An important exception to this scenario is the research at the Federal University of Paraná (Curitiba, Brazil) on molecular diagnosis of leishmaniasis and tuberculosis. In 2003 this research resulted in the development of a diagnostic kit for Leishmania (Viannia) braziliensis. After developing the technology at the university, the researchers transferred it to the Centre for Production and Research of Immunological Products (CPPI) (Curitiba, Brazil), for manufacturing and distribution. Leishmaniasis is spreading sharply in several areas of the world as a result of co-infection of HIV/AIDs and leishmaniasis. Leishmaniasis patients are highly susceptible to contract HIV and in HIV/AIDS patients the presence of leishmaniasis accelerates the onset of AIDS by cumulative immunosuppression and stimulation of virus replication. So far, 33 countries worldwide have reported co-infections. The leishmaniasis diagnostic kit is produced by CPPI at one-third of the cost of diagnostic kits for leishmaniasis in India. While the research is led by the Federal University of Paraná, the Tropical Disease Research Centre at Mahatma Gandhi Institute of Medical Sciences (Sevagram, India) has participated and contributed complementary expertise. Also the partners are working on further developing the tuberculosis kit and using their respective strengths in making the kit as cheap as possible.

7.5.3 Impacts of the entrepreneurial collaborations

The main impacts of the India-Brazil collaborations are increased market access of Indian firms in Brazil as well as in other Latin American countries. The Brazilian pharmaceutical market has been growing in recent years and is forecast to increase by a 7.1 per cent average annual growth rate, reaching a value of US $18.3 billion in 2012 (BMI, 2009). This steep growth rate has attracted firms from various countries, including developing countries, to Brazil and has led to their market expansion and increased economic revenues.

A significant impact of the India-Brazil health biotechnology collaboration has been increased availability of cost-effective health products. Indian health biotechnology firms have proven the ability in process innovation that has lowered the price of health products such as the hepatitis B vaccine. Brazilian firms can also contribute cost-effective health products to the Indian market. In Brazil, for example, AIDS and leishmaniasis diagnostic kits are available at prices 30–40 per cent less than the cost that they are in India (Box 7.4). By using research collaboration between the two countries, there is the potential—and desire—to make these technologies available to the public in a way that would increase accessibility through affordability.

7.6 Conclusions

The case-study research on India's health biotechnology collaborations with Bangladesh and Brazil has identified several findings that highlight the potential for South-South collaboration and shed light on approaches to strengthen it. The main messages from this research are:

South-South collaboration has become an integral part of India's foreign policy but its focus is narrow. India emphasises collaboration with developing countries for political reasons, to strengthen their voice in international fora and as a way of becoming less reliant on the North. Recent years have seen renewed emphasis on India's South-South collaboration, for example, with the establishment of the IBSA programme. In the health biotechnology field, India's collaboration with advanced developing countries has expanded, but its governmental emphasis on collaboration with the rest of the developing countries conveys a message that is not in tune with the wider foreign policy emphasis. There is a need to enhance more joint R&D activities generally in science and technology and, more specifically, in health biotechnology with larger Third World countries in order for India's South-South collaboration to flourish and have further impacts.

India's health biotechnology collaboration with developing countries has been more focussed on capacity building than joint research. As India's economy has strengthened in recent years, we have started to observe increased Indian contributions to capacity building in fellow developing countries. These are welcome developments and reflect the powerful potential of South-South learning in the health biotechnology field. India has been strengthening its capacity in health biotechnology and has started to share it with other developing countries. Still, the limited Indian R&D collaboration with developing countries shows how under-utilised the South-South collaboration is as a tool for addressing joint health problems. As India's economy grows, and the country allocates increased funding to research, it should put increased fillip to joint R&D with other developing countries to utilise their different strengths to address global health problems.

Indian research collaboration with select developing countries is increasing and shows promising potential. India and Bangladesh have been increasing their health biotechnology collaboration extensively in recent years. It has been focussed almost entirely on cholera, a significant health problem in Eastern India and Bangladesh. Through a combination of genomic sequencing within Bangladesh, and the transfer of technology to India, a new cholera vaccine was developed. As health biotechnology capacity in Bangladesh is limited, it is not surprising that India's collaboration has been with an international organisation, like the ICDDR,B. This reflects both that a minimal capacity is needed in both developing countries in order to engage in South-South collaboration, and how international organisations can provide the needed funding for it. Bangladesh has now made plans to promote its biotechnology sectors, and if India allocates more funding towards South-South research collaboration, their reliance on international organisations may diminish in the future.

India's South-South collaboration has expanded its market and increased the availability of affordable health products. Almost all the India-Brazil collaborations reviewed in this study are found to be focussed on gaining market access for Indian firms in Brazil. This has led to increased revenues for Indian firms and at the same time also increased availability of affordable health products in the Brazilian market. Still, the collaboration so far has been rather one-sided and the Brazilian side has not contributed much to it apart from their market. As the Brazilian entrepreneurial activities in the health biotechnology sector are strengthened, they can contribute more fully to the collaboration and, through it, gain access to the important Indian market.

Our study shows that India values South-South collaboration and has expressed the intention to increasingly invest in the collaboration with fellow developing countries. Both its bilateral and multilateral ties reflect renewed emphasis on collaboration with developing countries. There is a great scope for India to contribute to health biotechnology development in other countries and for all parties to gain mutual benefits provided other developing countries also see this and do not allow a narrow nationalist agenda to prevail. In this context, India also has to support its words with action in order for the collaborations to gain impetus and flourish. The next few years will show if India, with the aggregate increase in the size of the economy and related economic clout, will contribute to a new dawn of South-South collaboration or if it will further enhance the knowledge divide among the developing countries, much at the cost of Nehru's cosmopolitanism, which in all these years has emerged as key component of India's foreign policy.

References

Ablett, J., A. Baijal, E. Beinhocker, D. Farrell, A. Bose, U. Gersch, E. Greenberg, S. Gupta and S. Gupta (2007). *The 'bird of gold': The rise of India's consumer market*. San Francisco: McKinsey Global Institute. Available at *http://www.mckinsey.com/mgi/reports/pdfs/india_consumer_market/MGI_india_consumer_executive_summary.pdf*

Al Abbas, K.M. (2009). "Bangladesh-India agreements signed", *Ground Report*.

ANI (2009). "India welcomes FDI from Bangladesh", Bangkok: Thaindian News.

BMI (2009). "Brazil pharmaceuticals and healthcare report Q1 2008", *Buisness Monitor International*. Available at *http://217.114.165.229/reportinfo.asp?report_id=705613*

Chaturvedi, S. (2005). *Dynamics of biotechnology research and industry in India: Statistics, perspectives and key policy issues.*

———. (2008). *Emerging patterns in architecture for management of economic assistance and development cooperation: Implications and challenges for India.*

DST (2008). *Research & development statistics at a glance, 2007-2008*. New Delhi: Department of Science and Technology, Ministry of Science and Technology, Government of India. Available at *http://dst.gov.in/scientific-programme/r&d-eng.pdf*

———. (2009). *International S&T cooperation*. New Delhi: Department of Science and Technology, Ministry of Science and Technology, Government of India.

———. (2010). "International S&T Cooperation", New Delhi: Department of Science and Technology, Ministry of Science and Technology, Government of India.

Kumar, N.K., U. Quach, H. Thorsteinsdóttir, H. Somsekhar, A.S. Daar and P.A. Singer (2004). "Indian biotechnology—rapidly evoling and industry led", *Nature Biotechnology*: DC31-DC36.

Matthew, J.C. (2007). "Export barriers hit pharma firms", *Buisness Standard*. New Delhi.

MEA (2005). *Bandung: Celebrating the fifieth anniversay of the Asian-African conference, 1955*. New Delhi: Ministry of External Affairs, Government of India, External Publicity Division.

———. (2008). 'ITEC'. New Delhi: Ministry of External Affairs, Government of India.

MoF (2010). *Economic survey 2009-2010*. New Delhi: Ministry of Finance, Government of India. Available at *http://indiabudget.nic.in/es2009-10/esmain.htm*

Mokoena, R. (2007). "South-South co-operation: The case for IBSA", *South African Journal of International Affairs* 14(2): 125-145.

NAM & ST (2010). *Centre for science and technology of the non-aligned and other developing countries*.

OECD (2009). *Globalisations and the emerging economies: Brazil, Russia, India, Indonesia, China and South Africa*. Paris: Organisation for Economic Co-Operation and Development.

Padma, T.V. (2005). "Asian nations plan technological cooperation", *Scidevnet*.

Parthasarathi, A. (2000). "India's experience with TCDC", *Cooperation South* 1.

Sawahel, W. (2005). "India to host Asian biotech training centre", *Scidevnet* 10. November.

The World Bank (2008). *Public and private sector approaches to improving pharmaceutical quality in Bangladesh*.

Thorsteinsdóttir, H., A.S. Daar, P.A. Singer and E. Archambault (2006). "Health biotechnology publishing takes-off in developing countries", *International Journal of Biotechnology* 8(1): 23-42.

UNDP (2010). *Human development report 2010 - 20th anniversary: The real wealth of nations—Pathways to human development*. New York: United Nations Development Programme.

WHO (2005). "Regional strategic framework for elimination of kala-azar from the South-East Asia region (2005-2015)", *WHO project* no. IND CRD 714. New Dehli: World Health Organization.

8 Promoting an African Renaissance? South Africa's Health Biotechnology Collaboration with Sub-Saharan African Countries

AUTHORS: *Victor Konde, Sidar Abdusamad Andrew Kapoor, Halla Thorsteinsdóttir*

8.1 Introduction

South Africa has had, to say the least, a tumultuous history. Its government's policy of apartheid in 1948, severed historical ties with some of today's Southern African Development Community (SADC) countries as well as with the international community. The consequences of apartheid forced South Africa to turn inwards, and develop its own scientific and technological solutions to address its challenges in relative isolation from the international community. Following the abolishment of apartheid in 1994, South Africa opened up to the rest of the world and has taken a strong interest in promoting development in Africa. It is Africa's largest and strongest economy in terms of gross GDP and one of only five sub-Saharan countries to be classified among upper/middle-income countries by the World Bank (The World Bank, 2010), South Africa is also classified as an emerging economy by the OECD (OECD, 2009).

South Africa has emerged, according to UNCTAD, as one of the top investors in Africa. For example, South Africa's outward foreign direct investment (OFDI) flows in Africa grew from about US $71 million to US $4.365 billion between 1990 and 2004 (UNCTAD, 2005). Within SADC,

South Africa's OFDI stock in the other 13 members of SADC increased from about US $806 million to US $1.31 billion between 1993 and 1997 (UNCTAD, 1999). South African firms such as MTN Group (operates in 16 African countries), SABMiller plc (present in 13 African countries), and Shoprite Holdings (operates in 16 African countries), among others, are actively engaged not only in the SADC but in many parts of Central, Eastern, and Southern Africa. South Africa's strong role as an investor in Africa is not exactly surprising as South Africa accounts for about 22 per cent of Africa's GDP and has a GDP per capita of about US $5,600—about 3.5 times higher than that of Africa (UNCTAD, 2010). South Africa's desire to play a key role in Africa is also enshrined in its international relations. The country prioritises collaboration with the neighbouring South African Development Coordination Conference (SADCC) countries and other countries in Africa in its strategic plan (DIRC, 2009).

Given its high investment in education and R&D facilities, South Africa attracts students and researchers from the entire region. It is the only country in Africa with a university among the top 200 universities in the world (University of Cape Town at 146), according to a 2009 ranking (QS, 2009). Its population has a relatively high level of education, and one of the highest education indices, of all African countries (UNDP, 2010). As a result, South Africa is Africa's major producer of science, and its authors contribute 30 per cent of the continent's publications in international peer-reviewed journals (Pouris and Pouris, 2009). The main scientific fields in which the country publishes in the international peer-reviewed literature are clinical medicine and plant and animal sciences (Jeenah and Pouris, 2008). South Africa has also invested significantly in its biotechnology sector and publishes papers in international peer-reviewed journals in this field that are cited more frequently than other leading developing countries in health biotechnology (Thorsteinsdóttir *et al.,* 2006). The country's emphasis on biotechnology is relatively recent: it was only in 2001 that South Africa presented its biotechnology policy (DST, 2001). From the early days, South Africa has built its biotechnology endeavours on a solid foundation in life sciences research. Even though most of the biotechnology activities in South Africa are carried out by South Africa's universities and public research organisations (Motari *et al.,*

2004), industrial development in the sector has taken off (Al-Bader *et al.,* 2009).

South Africa has also started to take on a growing role in promoting international development (Braude *et al.,* 2008). Its international development activities are directed almost entirely towards other African countries, where South Africa has assumed an increasingly prominent leadership role in promoting political stability, trade, and scientific development. This is demonstrated by the creation of African cooperation units in a number of government ministries, including the Department of Science and Technology (DST). Given that South Africa has capacity in health biotechnology, and a political emphasis on collaborations with fellow countries in sub-Saharan Africa, there is scope for the country to both promote capacity building and collaborate in the health biotechnology sector. So far, though, there is a lack of research on the extent to which South Africa has worked with sub-Saharan countries in extending capacity and promoting innovation in health biotechnology. To address this gap, we carried out case-study research on South Africa's health biotechnology collaboration with sub-Saharan countries (see Box 8.1).

To better understand the wider context of South Africa's health biotechnology collaboration, we delineate the main initiatives and policies in science and technology that the South African government has built with fellow developing countries. We also map the key research and entrepreneurial collaborations South Africa has with other developing countries. Then, we focus in on specific collaboration initiatives, and examine the driving forces behind South Africa's collaborative efforts in health biotechnology, the challenges they have faced, and the impacts the collaborations have had so far. We carry out a similar assessment in two African countries, Kenya and Zambia, to determine their policies and perceptions and what drives collaboration with South African institutions in health biotechnology.

BOX 8.1

Case-study research on South Africa's collaborations: Methodology

To examine South Africa's collaborations with sub-Saharan countries, we chose to focus on the country's collaborations with Kenya and Zambia. We studied a total of two cases and defined each case as the bilateral collaboration between South Africa and Kenya *versus* South Africa and Zambia. Kenya is relatively advanced among sub-Saharan countries in terms of science and technology development, but Zambia has limited scientific capacity. Both countries see the biotechnology sector as being of high importance and have developed biotechnology strategies. Zambia is in the SADC region, while Kenya is not in the same regional economic community (REC) with South Africa. However, Zambia belongs to two RECs: SADC and, together with Kenya, in COMESA. As such, Zambia is in one REC that includes Kenya, and in another REC that includes South Africa. Therefore, we could examine regional science and technology cooperation within different RECs.

To understand the potential of South Africa's collaboration with sub-Saharan African nations, we conducted case-study research with multiple sources of data. We relied heavily on interviews with key experts in South Africa, Kenya and Zambia about their views and experiences of South-South collaborations in health biotechnology and interviewed 48 experts for this study (Table 8.1). We combined this with input from our policy analysis, scientometric analysis, and an examination of the firm survey data, to gain evidence-based insight into South Africa's collaboration with sub-Saharan countries.

TABLE 8.1

Breakdown of number of interviewees in case-study countries

Countries	Number of interviewees
South Africa	27
Kenya	10
Zambia	11
Total	48

8.2 Governmental interest and support

South Africa's desire to play a key role in Africa is reflected in its international relations. The country seeks to 'continue prioritisation of the African continent, strengthen political and economic integration of the SADC, strengthen South-South relations, strengthen relations with strategic formation of the North, strengthen political and economic relations and participate in the Global System of Governance' (DIRC, 2009). The government in South Africa has placed great emphasis on close ties with other African countries. It has a vision of an African renaissance, reflected in the 2001 establishment of the so-called African Renaissance and International Co-operation Fund. The spectrum of initiatives that the fund supports is wide, and there is a strong focus on supporting peacekeeping, strengthening democratic institutions, and socioeconomic advancement. While capacity building in other African states is a part of South Africa's collaboration efforts, the emphasis on supporting R&D seems to be small. South Africa places priority on the neighbouring SADC member states in their collaborative efforts, and these countries receive the bulk of support.

South Africa's biotechnology strategy puts an emphasis on international collaboration. It recognises the need to strengthen international networks to improve knowledge generation and technology acquisition. It also promotes actions such as stronger international linkages with equivalent institutions that are at the frontier of biotechnology development: joint projects with equipment suppliers; a stronger role for biotechnology in bilateral agreements; and biotechnology as a cornerstone of the Millennium Africa programme to promote African development (DST, 2001). There is no explicit mention of South-South collaboration in the strategy, but it is implicitly included.

South Africa also channels significant portions of its African support through multilateral institutions such as the New Partnership for Africa's Development (NEPAD), which has become part of the African Union's Science and Technology Consolidated Plan of Action. NEPAD's mandate is to develop an integrated socioeconomic development framework throughout the African continent, and the headquarters of NEPAD's Science and Technology programme is in South Africa. NEPAD has an explicit emphasis on cross-border cooperation in Africa, and biotechnology is one of relatively few technological areas prioritised by NEPAD. For instance, it established the Bioscience Initiative in 2005, which includes the formation of regional networks in Africa's five main regions: Central, Eastern, Northern, Western and Southern. Based in South Africa, the network is called Southern African Network for Biosciences (SANBio), whose mandate includes delivery of health solutions based on biotechnology knowledge for the health needs of the region, such as HIV/AIDS, malaria and tuberculosis (Makinde *et al.,* 2009). Through NEPAD, thus, we observe a strong emphasis on biotechnology as a priority area for cross-African collaborations.

South Africa has also signed a number of bilateral agreements to support scientific and technological collaboration under the auspices of its DST. The DST now has such agreements with a relatively large number of developing countries. Between 1994 and 2006, DST signed formal agreements with 23 countries (NRF, 2010). Most of the active agreements seem to be with developed countries or the emerging economies. But DST has also developed a five point bilateral strategy: '1. To support NEPAD and SADC; 2. Market South African R&D services and products; 3. Create opportunities for joint projects; 4. Share expertise and resources with less developed countries; and 5. Position South Africa strategically within emerging high growth markets in Africa' (DST, 2009). The primary focus of this strategy seems to be the promotion of South-South collaborations within Africa, and to share South Africa's scientific wealth. Notably, the bilateral strategy also includes an emphasis on marketing. So far, South Africa has established bilateral relations in S&T with at least 13 African countries.

In biotechnology, South Africa is also an active member and host of the International Centre for Genetic Engineering and Biotechnology (ICGEB), an international network. ICGEB is a part of the UN system with three main components (in Trieste, Italy; Delhi, India; and Cape Town, South Africa) and 39 affiliated institutions (one in each country) that form a South-South network. The Cape Town component was inaugurated in 2007 and intends to play a significant role in research and training activities for improving biotechnology R&D both in infectious and chronic diseases that affect Africa. It will also focus on plant biotechnology research and training to improve the staple crops essential for the African population.

Other major collaborative efforts emphasised by the government of South Africa include the IBSA trilateral agreement, whose main objectives include the promotion of *'South-South dialogue, cooperation, and common position on issues of international importance between the three countries'* (Mokoena, 2007). As discussed in Chapter 4, the three countries decided to focus their S&T cooperation in six specific fields, including biotechnology, HIV/AIDS, malaria and tuberculosis. South Africa leads the work in biotechnology and tuberculosis.

From this outline, it is clear that there is a significant will for South-South collaboration by the South African government, and that South Africa has S&T collaborations with a wide variety of countries. The varied linkages with other developing countries are recent, having been established over the last 15 years. Politically, there is a strong emphasis on collaboration with other African countries, particularly with the SADC countries, or as a part of the NEPAD framework. Still, South Africa's South-South focus is not exclusively on Africa, and both bilateral and multilateral initiatives are now also emphasising collaboration with other Southern countries, particularly with the emerging economies.

8.3 The geography of South Africa's health biotechnology collaboration

8.3.1 Mapping research collaboration

To examine the level and distribution of South Africa's South-South collaboration in health biotechnology, we mapped its research and entrepreneurial collaborations (see Chapter 1 for details of the methodology). We identified 378 papers co-authored by researchers from South Africa with at least one other developing country from the SCOPUS database for the period of 1996 to 2009. The collaboration level is small compared to leading developing countries in South-South health biotechnology collaboration, such as Brazil, which published over 1,000 papers in South-South collaboration during the same time period. Before 2002 South Africa published almost no health biotechnology papers with other developing countries, but since then has increased its publishing rate from around 20 to over 60 papers in 2008.

The results of mapping South Africa's research collaboration show that Brazil and China are the main countries South Africa collaborated with (Figure 8.1), nevertheless, there were only around 60 co-authored papers with these countries during the 15-year period we studied. India comes in third place, with around 40 co-authored papers for the period studied. It is noteworthy that only four sub-Saharan countries are on South Africa's top-10 list of main health biotechnology research collaborators. Kenya is the sub-Saharan country that collaborated most heavily with South Africa, and shares the fifth place overall with Mexico. The low level of collaborations between South Africa and sub-Saharan countries, particularly SADC countries, was confirmed by an analysis of scientific collaboration in general, where only 3 per cent of papers produced by SADC were jointly authored by those countries (Boshoff, 2010). Most of South Africa's collaborations are with high-income countries. Considering that the number of health biotechnology papers South Africa co-authors with developing countries is low, a single collaboration initiative that is productive, in terms of leading to several co-publications, could change the research collaboration pattern we observe. Our analysis of research collaboration in health biotechnology is thus more accurate in measuring the linkages in the past than in predicting future collaborations. Overall, the mapping of South Africa's research collaborations with other low- and middle-income countries does not suggest that South Africa has been forming ties with other sub-Saharan countries in health biotechnology. Kenya, Nigeria, Zimbabwe and Uganda may be the exceptions.

We looked at the subfields of health biotechnology of South Africa's South-South collaboration. Just like in the other developing countries 'Genetics and Heredity' is the most common subfield of South Africa's collaboration. 'Virology' is then in second place and 'Immunology' is in third place. This pattern is different from the patterns of the other focal countries in this study where traditional biotechnology fields such as 'Biochemistry and Molecular Biology' and 'Microbiology' typically were ranked second or third. The emphasis on 'Virology' and immunology can reflect significant South-South collaboration in examining HIV/AIDs, a serious health problem in South Africa and many other sub-Saharan Africa countries and working on a vaccine against the disease. 'Tropical Medicine' ranked in seventh place as a subfield for South Africa's South-

South collaboration reflecting a relative emphasis on diseases of the African continent.

FIGURE 8.1

The main developing countries South Africa collaborates with in South-South research collaboration in health biotechnology, based on the number of co-authored papers between 1996–2009

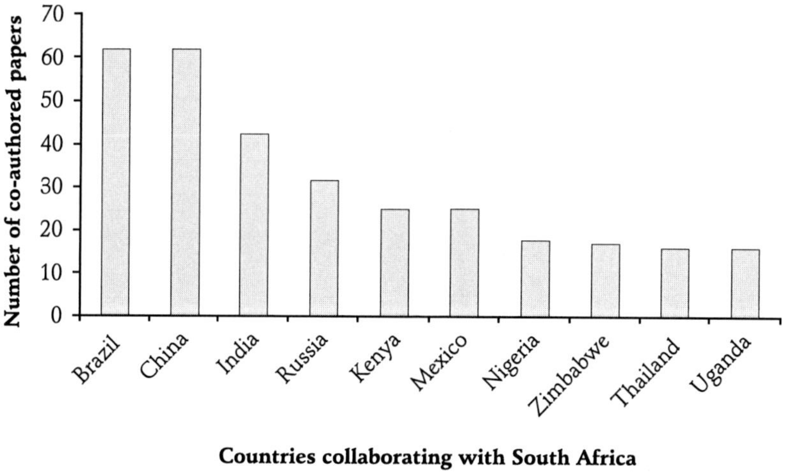

Source: Authors' presentation of data compiled by Science-Metrix using Scopus.

We also examined to what extent South Africa was engaged in South-North collaboration. It co-authored around 1,700 health biotechnology papers from 1996 to 2009 with authors from high-income countries, increasing its publication rate from around 100 papers a year in the 2002, to 200 papers in 2008. South Africa's main Northern collaborators are the United States (731 papers), United Kingdom (522 papers), Germany (224 papers), and France (205 papers). It is noteworthy how commonly South Africa collaborates with authors from the United Kingdom, reflecting historically strong ties between the two countries.

8.3.2 Mapping entrepreneurial collaboration

When we map the developing countries that South Africa has at least three entrepreneurial collaborations with (see Chapter 1 for details of the methodology), we see different links from those that involve research collaborations (Figure 8.2). India is by far South Africa's main entrepreneurial collaborator in the health biotechnology field. What is striking is that South Africa has relatively greater emphasis on entrepreneurial ties with other sub-Saharan countries than research ties. Botswana, Malawi, Nigeria, Namibia and Zimbabwe all rank amongst South Africa's main entrepreneurial collaborators in the health biotechnology field.

FIGURE 8.2

The main developing countries South Africa collaborates with in South-South entrepreneurial collaborations in health biotechnology based on a survey of firms[1]

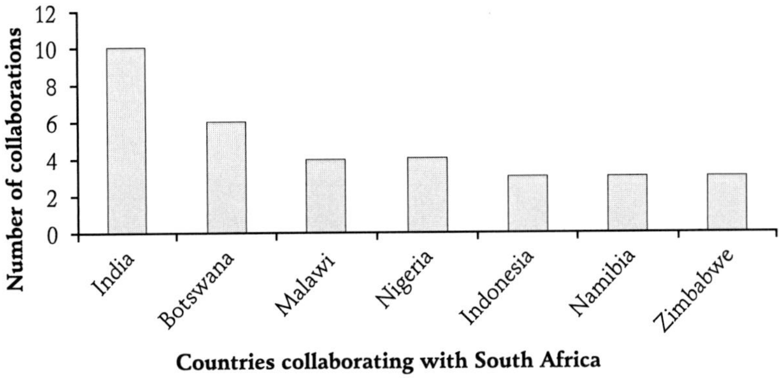

Source: Authors' presentation of data compiled by own survey.

From mapping South Africa's collaboration linkages we get conflicting messages on the geography on South Africa's past health biotechnology collaborations. Its research collaborations seem to have emphasised some of the emerging economies, whereas entrepreneurial collaboration has a stronger emphasis on sub-Saharan Africa. Both types of collaborations are rather modest, so the pattern we observe from looking at past collaborations could easily change in the future. Because of the political emphasis on African collaborations, we decided to focus further analysis on South Africa's collaboration with Kenya and Zambia. Kenya is a relatively advanced sub-Saharan country in terms of science and technology development, but Zambia has limited scientific capacity. For instance, while South Africa published around 3,500 papers in health biotechnology between 1996 and 2009, Kenya published over 600 papers,

1. Note the figure shows only countries that South Africa has at least three collaborations with.

and Zambia only 84 papers. All three countries are Anglophone, and thus likely to be members of common science and technology networks and bodies.

There is also significant international trade between these countries in pharmaceutical products. Zambia, Kenya, Zimbabwe and Ghana (in descending order) were the top four export markets for South African pharmaceutical products in 2009, according to UN Commodity Trade Statistics Database (Comtrade) database (COMTRADE 2010). In general, Zambia and Kenya accounted for about 33 per cent of South Africa's exports of pharmaceutical products. Likewise, South Africa and Kenya accounted for about 10 per cent and 5 per cent of Zambia's exports of such products in 2009.

Kenya and Zambia have both singled out the biotechnology sector as being of high importance, and have developed biotechnology strategies. In 2006, the Kenyan government approved the National Biotechnology Development Policy (Ogodo, 2006). The policy identifies industry and trade as key areas for Kenya's use of biotechnology, offering the country a positive—if limited—opening into the biotechnology sector. It also notes the benefits that come from the safe development and application of biotechnology in agriculture, environment, health, industry and research. Even though some references are made to health biotechnology, the country's primary focus in the biotechnology sector remains on agriculture (Republic of Kenya, 2006). Harnessing international collaboration is a part of Kenya's overall biotechnology strategy, but no reference is made to collaborations being South-South as opposed to South-North. If we look at Kenya, its emphasis is much stronger on South-North than South-South collaboration. It co-authored 527 health biotechnology papers with authors in high-income countries *versus* 144 papers with authors in other low- and middle-income countries.

Similarly, in 2003, Zambia honed in on the biotechnology sector with its National Biotechnology and Biosafety Policy. It emphasised capacity building for agricultural regulations in the country, particularly to hinder unconstrained dissemination of genetically modified organisms in the country and to protect Zambia's crop diversity (Government of Zambia, 2003; Hanyona, 2003). This strong focus on agricultural biosafety is related to the decision by the Zambian government to reject donations of genetically modified maize from the United States because of concerns about the safety of ingesting genetically modified organisms during the 2002 drought that afflicted most of Southern Africa (Pascal Newbourne Mwale, 2006). This debate has almost disappeared as production of food crops has more than doubled since then. For example total cereal production increased from 754,966 metric tonnes to 1,630,173 metric tonnes between 2002 and 2008, according to FAO (FAO, 2010).

There is limited discussion of health biotechnology in Zambia's biotechnology policy, but some R&D is aimed at protecting the population against preventable diseases and developing capacity to diagnose tuberculosis, HIV/AIDS and malaria. The biotechnology policy only makes mention of international collaborations in reference to human resource development. The report states that, 'Zambia shall rationalize its investment by making maximum use of existing regional, international and other educational and training bodies in biotechnology and biosafety, and by preparing its undergraduate students for easy entry into such programmes by means of curriculum stream options.' (Government of Zambia, 2003). Zambia is engaging in both North and South collaboration in the health biotechnology field. It has co-authored 77 papers in health biotechnology with authors from high-income countries *versus* 36 papers with authors in other low- and middle-income countries.

Thus, both Kenya and Zambia have recently put an emphasis on biotechnology. Their plans are oriented more towards agricultural biotechnology rather than health, but there is still some mention of the importance of building capacity in health biotechnology. As South Africa is the most advanced African country in health biotechnology, there is scope for the country to provide capacity and research in the field in sub-Saharan Africa. In what follows, we examine South African efforts in this respect.

8.4 Collaboration in research activities

To gain a better understanding of South Africa-sub-Saharan Africa research collaborations, we interviewed researchers in South Africa, Kenya and Zambia. We talked to the heads of some of the major research institutions and researchers in the three countries who had co-authored papers with each other. Both Kenya and Zambia have leading medical research institutes. The Kenya Medical Research Institute (KEMRI, Nairobi, Kenya) is a state corporation, established in 1979 as the main national organisation responsible for carrying out health research in Kenya. It collaborates with some of the leading organisations in health research globally such as the British Medical Research Council (MRC) and the Centers for Disease Control and Prevention (CDC) in the United States. Some of its regional collaborators include the Blair Research Institute (Zimbabwe), Ethiopian Health and Nutrition Research Institute (Ethiopia), Makerere University School of Medicine (Uganda), Medical Research Institute (South Africa), the National Institute of Medical Research (Tanzania), Noguchi Memorial Institute for Medical Research (Ghana), and the University of Zambia School of Medicine (Zambia).

The Tropical Diseases Research Centre (TDRC) (Ndola, Zambia) was established in 1975 jointly by the WHO and the Zambian Government. It is a national resource dedicated to research and training to address priority health problems in Zambia. Its main collaborators are Northern institutions such as Boston University (United States), the Institute of Tropical Medicine (Belgium), the University of Alabama (United States), and the London School of Hygiene and Tropical Medicine (United Kingdom). Some of its regional collaborators include the University of KwaZulu-Natal (South Africa), the University of Ghana (Ghana), the University of Nairobi (Kenya), and The Blair Institute and the University of Zimbabwe (Zimbabwe). In the next sections we outline the reasons, challenges, and impacts that collaborators are faced with in the context of South African research collaborations.

8.4.1 Reasons for the collaboration

Analysis of the interview results reveals several motivations for collaboration in South Africa. The main ones stressed in the interviews are described below.

1. *Capacity building:* When interviewing South African researchers they stressed capacity building as a strong motivation for collaborations with Kenya and Zambia (see for example Box 8.2).

BOX 8.2

Growing African capacity

Within Africa, South African researchers are playing a vital role in capacity building, technological transfer, and knowledge exchange. In these efforts they are using what they are learning through their own development and through collaboration with other more scientifically advanced nations.

For example, researchers from the University of Stellenbosch (Tygerberg, South Africa) led an International Project, funded by partners such as the International Atomic Energy Agency (IAEA), to transfer molecular technologies to various countries in Africa. This training programme allows researchers from over 10 African nations, including countries such as Ghana, Kenya, Sudan, Tanzania and Zambia, to come and gain training in South Africa. The programme focusses on building research capacity and developing transfer skills using tuberculosis as a relevant local model to transfer molecular biology techniques, technologies, and know-how to visiting researchers from other African institutions. Visiting researchers are able to work on samples from their home countries while advancing their scientific skills useful for research on other endemic diseases such as HIV and malaria. Complementary to the basic lab skills, the programme also provides training in scientific communications, grant preparation, and writing for publication.

However, one of the main challenges facing research capacity building and exchange programmes in sub-Saharan Africa is the inability of researchers to return to their home countries and apply their newly developed skills and knowledge. This difficulty lies in the inability of other sub-Saharan African countries to integrate the new capacity of returning scientists into their innovation systems. Programmes such as the one pioneered by the ICGEB in Cape Town, have attempted to address this by providing researchers trained in South Africa a returning research grant in order for them to continue carrying out their research upon return to their home countries. The ICGEB maintains collaboration with the researchers monitoring their progress, helping them start labs, build on their projects, and provide additional assistance where required to ensure that capacity is being built and translated directly to the other side.

Commonly in sub-Saharan Africa, the benefits of such training tend to be minimised when there are not appropriate receptors in the country of origin to channel the new capacity and skills. The ICGEB model can be an effective way to ensure that the first step in overcoming the critical barrier limiting uptake into the African innovation system is minimised.

The capacity building was focussed on strengthening both research capacity in Kenya and Zambia as well as capacity for clinical trials. The message was that the South African contribution to capacity building was well aligned with needs in other sub-Saharan countries, and provides affordable solutions. A South African researcher commented, for instance: 'First of all we developed the methods and then we transfer it. Oh yeah, we develop them. And what we try to develop, because we are also a third world country, the methods that we develop must be cheap and user friendly, mustn't be high tech technology.' The ICGEB centre is an example of capacity-building efforts in this field. It is based on both international and South African expertise that is being extended to sub-Saharan countries, including Kenya and Zambia. Even though many African countries were members of ICGEB, prior to the establishment of the South African component, their nationals were not taking full advantage of the training potential of ICGEB. This has changed dramatically since the establishment of Cape Town Component funded mainly by the South African Ministry of Science and Technology.

Several of the interviewees mentioned that there was a need for Africans to work together to address their problems. Even though South Africans are generally the providers of technologies and specialised health biotechnology knowledge, they also gain from the collaboration. One South African researcher stressed that they could not

> close their doors to the problems of their neighbouring countries. Instead we need to take our duty as citizens of Africa and increase the capacity to our mutual benefit... It is not a one-way situation, there are contributions that other countries make as well. It may not be a financial contribution it may be an intellectual contribution, or resources, natural resources contribution as well.

2. *Shared health problems:* South Africa, Kenya and Zambia share a high burden in infectious diseases, such as HIV/AIDs and tuberculosis. As a result, the researchers we interviewed commonly cited shared health problems as a driver for research collaboration. The governments of all three countries prioritise these areas

and their biotechnology strategies make reference to building up biotechnologies that can be applied to address these health problems. For instance, the South African biotechnology strategy states: 'Biotechnology can make an important contribution to our national priorities, particularly in the area of human health (including HIV/AIDS, malaria and TB)' (DST, 2001).

Several formal networks encourage research on shared health needs in Africa. Examples of such networks are the African Malaria Network Trust (AMANET), the African Network for Drugs and Diagnostics Innovation (ANDI) (Box 8.3), and the Afro-Immuno Assay network (AIA). They aim to encourage local capacity building and research collaboration to address shared health needs and reflect intertwined drivers of shared health problems and capacity building in sub-Saharan collaboration. Both the TDRC in Zambia and KEMRI in Kenya are among AMNET-supported or participating centres.

South Africa and Kenya carry out substantial levels of laboratory-based discovery research in health biotechnology aimed at addressing their health needs, whereas Zambia seems to focus on surveillance and clinical trials. This reflects a realisation that their limited resources are best used to manage diseases rather than attempt to develop new-to-the-world innovation. By collaborating with other countries in Africa, they can jointly contribute to developing new health products aimed at addressing local health needs.

3. *Exploiting strategic differences in resources:* Here, the meaning of resources ranges from centres located in areas with key differences in infection rates, genotypes of parasites, and environment, to those with key R&D resources and expertise to undertake research. In this regard, many scientists we interviewed worked with South African institutions or scientists to learn methods. For example, Macha Malaria Research Institute in Zambia worked with South Africa's Medical Research Institute and Witwatersrand University (Johannesburg) to learn how to identify and classify mosquitoes morphologically and genetically. To achieve this goal, they sent

samples and graduate students to their South African partners. Since then, they have built up significant capacity not only to identify, but also to build their own colonies. Macha is now training the National Malaria Control Centre (NMCC) of Zambia to establish its own malaria colonies, and with the University of Zambia's School of Veterinary Medicine is working to establish an animal colony for feeding their colonies of mosquitoes. Other African partners of Macha include the Blair Research Institute in Zimbabwe and Zululand University in South Africa.

Similarly, a university-based researcher in South Africa described a project being conducted collaboratively with a hospital in Zambia. Zambia was chosen as a partner because of the country's high incidence of the type of malaria being researched, which made it easy to find subjects for the study. In order to conduct the research, however, they needed to build greater capacity. The hospital did not have the necessary equipment, nor did they have access to the technology available at the research centre in South Africa. Thus, in order to have access to a strain of malaria in Zambia, the South African partner brought equipment to the hospital in Zambia, and other supplies needed to build up the lab. They also brought in an expert lab technician who taught Zambian researchers new technologies, and more efficient ways of testing. The interviewee stressed that the partnership was beneficial for both countries; the capacity building happened in South Africa as well, in the form of cultural knowledge associated with the disease.

Others pointed out that participation and seeking partners in multi-centre research is important. First, multi-centre studies bring together partners in different environments, experiences and skills. For example, a KEMRI scientist said that their interest in participating in a multi-centre study was to learn new technologies and to understand immunological responses to natural infections by different age groups, different populations of people, and different genotypes of parasites that may not be found in one country. It influenced the choice of collaborating partners, especially in vaccine clinical trials, where it is necessary to understand how the vaccine works in different people who may be infected by a different type of parasite.

4. *Donor push:* A fourth reason for collaboration that we came across in our interviews was a push from donors or other Northern-based agencies. Donors may decide to support research in Africa, but will require the project to involve collaboration between two or more African countries. Their justification is to encourage resource-sharing and joint regional efforts to deal with African problems. The risk is that the collaboration may only be set up in order to access research grants and to fulfil donor requirements, rather than because researchers genuinely wish to work together. It can, thus, be collaboration in name only, or a fund-driven collaboration that will dissolve as soon as the grant is over. Also, in those cases the agendas of others—specifically of funding bodies—serve as the main reason for research collaboration.

   In other cases, South-South collaboration driven by Northern forces can be an efficient way to gain access to training and research expertise. For instance, the International AIDS Vaccine Initiative (IAVI) supports the efforts of several groups in South Africa to train researchers and clinicians from Kenya and Uganda in laboratory techniques and quality control of clinical trials (Stevens *et al.,* 2009). Such groups in South Africa include the University of Witswatersrand (Johannesburg, South Africa) and Contract Laboratory Services (Johannesburg, South Africa). South Africa can provide cost-effective training, more appropriate to local conditions than Northern-based training would offer. Shared experience and the close proximity of collaborators in Africa makes the training highly effective.

8.4.2 Challenges for the collaborations

The biggest challenge facing South Africa's collaborations with other sub-Saharan countries is the lack of resources available to fund these collaborations. Even though South Africa is placing an emphasis on investing in health biotechnology collaboration, and perceives collaboration with other sub-Saharan countries to be important, it is still a developing country with conflicting demands on its limited resources. In order for South Africa to be able to allocate resources towards capacity-building efforts in poorer countries to a significant level, the country needs help from international sources. The IAVI example described above is such a case, where South Africa can share its health biotechnology strengths with other sub-Saharan countries, partly because IAVI provides some of the funds. In our research, we identified other examples of outside organisations helping South Africa fund its capacity-building efforts in sub-Saharan countries. The WHO, for example, provides the University of Witwatersrand with funding to provide training to students from Africa on research on mosquito vector control for malaria; and the IAEA supports technology transfer activities by the University of Stellenbosch for tuberculosis diagnosis to researchers working in rural African environments. Researchers we interviewed for this study described the importance of these contributions by international organisations. This view was expressed by a malaria researcher who said:

> South Africa is financially well off in the bigger picture. But even the funding we get from South Africa is, if I don't get international funding half the students here would go home. So we cannot do what we do without international funding. We use local funding to supplement international funding but it can't be the other way around.

Another challenge to research collaborations that was emphasised by the interviewees was that it was sometimes difficult to keep up the momentum in the long term. They said that what happens in many cases is that students or other research personnel come to South Africa to train or to initiate research projects. While in South Africa, they are actively engaged in research and the collaboration flourishes, but when they return to their home countries in sub-Saharan Africa, the collaboration fizzles out. A typical scenario is that the researcher returning home has such a heavy teaching load that he/she cannot allocate enough time for research. Also, they may not have the research infrastructure in their home countries, or enough funding for research expenses to continue the work. The following statement by one interviewee illustrates this challenge: 'In my experience, once they have left here, with all the promises, nothing will happen, and then it comes back to a standstill when, once they are back into their country. Because they are overwhelmed with all kind of other things.'

It is a waste of resources to train people to do research that is highly relevant to health needs in their countries, because they do not have time or facilities to do research upon their return, and as such cannot give back to their home countries. As a South African researcher stated: 'It doesn't help just to pour money into a bottomless pit and say we've got to train people. There must be a part contribution and commitment by the government of the country that is being assisted.' The ICGEB in Cape Town has come up with a strategy for remedying this challenge. They offer a three-year post-doctoral fellowship focussed on genomics/biotechnology research training. After the fellows complete their training and return to their home countries, they receive a returning research grant. This makes it easier for them to continue carrying out research upon their return, and convenient for ICGEB to monitor the local capacity building in the recipient countries.

8.4.3 Impacts of the collaborations

The main impact of South Africa's research collaboration with Kenya and Zambia has been increased capacity building. Still, the level of South Africa's research collaboration with sub-Saharan African countries has been low. The capacity-building efforts have not enhanced health biotechnology competence in sub-Saharan Africa in any major way, but the experience has demonstrated that this model can work. According to an interviewee who has actively promoted capacity building all over Africa, 'some of them really have made progress and apply now these techniques for various kinds of other diseases.' Investment in people is of paramount importance, and when governments align their meagre research investments to leverage this capacity building then we can start to observe significant impacts from the collaboration. Increased local capacity in cutting-edge fields such as health biotechnology also has the result that researchers in poorer countries can reduce their reliance upon Northern countries.

Another important impact of South African collaboration with other sub-Saharan countries is the increased number of publications in high-impact journals that the collaborations have generated. As mentioned above, papers from South African health biotechnology researchers in general are cited more frequently than the papers of researchers from other leading developing countries, which leads to higher visibility for their research. A possible impact of sub-Saharan countries' collaboration with South Africa is, thus, higher visibility than would come from papers published solely from local research efforts. Still, considering that most sub-Saharan African countries publish much more in South-North rather than South-South research collaborations, and research by developed countries' researchers have considerably higher citation rates, South African citation rates are not likely to be a motivation for Kenya's and Zambia's sub-Saharan collaboration. To date, we have not seen examples of South Africa's health biotechnology research collaborations having an impact on the development of new products or services. Considering that South Africa's biotechnology sector is still nascent, it is not surprising that its South-South research collaborations have had limited influence on the development of new products and processes. To explore the potential impacts of South Africa's health biotechnology collaborations on African innovation further, we shift attention and analyse South Africa's entrepreneurial collaborations.

8.5 Collaboration in entrepreneurial activities

Mapping South Africa's entrepreneurial collaborations in health biotechnology shows that firms are relatively active in collaborations with other low- and middle-income countries. The private biotechnology sector in South Africa is young and small (Al-Bader *et al.,* 2009), but according to our research it still has established collaborations with other developing countries (Chapter 3). The linkages are with other emerging economies—particularly India—but also with other countries in Africa. However, when we looked specifically for collaborations between South African firms and Kenyan or Zambian firms, we did not identify much activity. The latter countries cannot be said to have real private sectors in the health biotechnology field, so we looked for collaborations involving the pharmaceutical sectors in those countries. In Zambia the technological base of the pharmaceutical firms was poor. According to a CEO in Zambia, biotechnology production is beyond their capacity: 'We are barely meeting the international standards on pharmaceutical products. Biotechnology products may add on another level of complexity.' This is similar in Kenya, where local manufacturers of pharmaceuticals are in chemical-based products rather than genomics or biotechnology-related products. As such, we did not have rich interview data on South Africa's firm collaborations, and thus we are basing most of the discussion below on survey data from collaborations between pharmaceutical firms.

The main entrepreneurial collaboration we could identify in the region was a network of firms around Aspen Pharmacare (Durban, South Africa) and Shelys Pharmaceuticals (Dar es Salaam, Tanzania). Shelys has production and sales agreements with three pharmaceutical firms in Zambia, one in Malawi, and three in Mozambique, as well as affiliates in Kenya and the Democratic Republic of the Congo (Congo-Kinshasa). In turn, in 2008 Aspen bought a 60 per cent stake of Shelys and licenses some of its products to the company. This enables Aspen products to enter Eastern and Central African markets. South Africa is thus collaborating with Tanzania, and through Tanzania with Congo-Kinshasa, Kenya, Malawi, Mozambique and Zambia.

8.5.1 Reasons for entrepreneurial collaboration

Our study indicates that the main reason for South African firms' South-South collaborations is market access. Almost all the firms that responded to our survey mentioned access to markets as their reason for collaboration. This certainly applies to South Africa's collaborations with other countries in Africa, which seem to be based almost entirely upon market access. In this vein, Aspen Pharmacare stresses that its alliance with Shelys Pharmaceuticals gave them access to the latter's distribution network in Africa and laid the foundation for its expansion into the Eastern and Central African markets (Aspen, 2009). The emphasis on marketing relations was also echoed by a Zambian CEO who stated: 'We will collaborate with anyone who has a good product that we can produce and market successfully.'

In general, providing knowledge or gaining knowledge were rare reasons for South African collaborations. The respondents to our survey only mentioned 'providing knowledge' once as a reason for their collaboration with the rest of Africa. There were, however, some instances of collaborations involving R&D between South Africa and Cuba. The Biovac Institute (Cape Town, South Africa) relies strategically on international partnerships for gaining access to knowledge, and collaborates with Heber Biotec in Havana (as well as with others) in order to transfer the technology to produce hepatitis B vaccine from Cuba to the Biovac Institute (Al-Bader *et al.,* 2009; Biovac, 2010).

According to interviews, there is evidence that the government, financial firms, and incubators responsible for biotechnology development in South Africa are the major drivers of the collaborations. A number of missions and trade arrangements were undertaken to promote market-based collaborations. These are primarily agreements with other African countries and some Latin American countries. The main objective from the recipient country appears to be infrastructure development.

8.5.2 Challenges for entrepreneurial collaboration

The main challenge for South Africa's entrepreneurial collaborations with the rest of Africa is in setting up the initial health biotechnology collaborations. South Africa has sent missions to set up collaborations in health biotechnology with other African countries, but the results so far have been minimal. Part of the problem is that firms in the target countries lack research, development, and production capacities. They may be able to base collaboration around simple pharmaceutical processes, but lack capacities in biotechnology. South Africa itself has a nascent biotechnology sector, and needs international collaboration for its own capacity building. As such, it is not yet likely to be in a position to expand capacity-building efforts to firms in other parts of Africa.

Poor communication, transport, and energy infrastructure were also mentioned in our interviews as limiting South African entrepreneurial collaboration with the rest of Africa. Unstable energy supplies, for instance, will dampen the interest of South African firms for setting up collaborative operations in other African countries. More importantly, South Africa has a very small range of biopharmaceutical products, unlike chemical-based drugs, that it can license to other African countries' firms. Most inter-firm transfers between relatively more developed countries and less-developed partners is likely to occur at the marketing phase than at R&D phases, as evidenced from the survey of the South African firms. We did not find any collaboration in R&D around traditional pharmaceutical

BOX 8.3

Formal networks of collaboration: Promoting African-led innovation

Several formal networks have been established in order to extend capacity in health research throughout the African continent. ANDI and AMANET are among the most prominent networks created to focus on health innovation for development in Africa.

The most recently developed network in Africa is ANDI which, was launched in Abuja, Nigeria in 2008. Led by a task force of prominent African researchers, health policymakers, and donor agencies, the network was born out of a realisation that novel approaches to supporting health product access, as well as R&D, are urgently needed in Africa (Mboya-Okeyo *et al.,* 2009). It specifically is aimed at boosting Africa's traditional lack of intra-continental collaboration in the field of health and biomedical research. The task force created a specific mission for ANDI which states that it will 'promote and sustain African-led health product innovation to address African public health needs through efficient use of local knowledge, assembly of research networks, and building of capacity to support economic development' (Nwaka *et al.,* 2010). Though the creation of ANDI is new, its early activities will centre around providing funding to enable networking between scientists, building a sustainable research environment, and promoting mechanisms to help scientists translate their research into products.

Another important formalised network influencing health innovation in Africa is AMANET. It has its origins in the African Malaria Vaccine Testing Network (AMVTN), which was established in 1995. AMANET was set up to build capacity and promote African led R&D specifically in the area of malaria. It has trained more than 900 African malaria researchers in Africa, established a network of trial sites linked to leading African institutions, and focussed on the strengthening of African malaria research including the development and testing of new malaria vaccine candidates (Kilama *et al.,* 2007). AMANET arose from the domination of a select few malaria research centres in Africa with strong traditional linkages with countries in the North. AMANET was thus created as a capacity-building network for African institutions in partnership, with other institutions in Africa and the North, to level the playing field in malaria research and increase Africa's participation in the research and clinical trial agenda. Strengthening activities

contd...

contd...

undertaken by AMANET include training workshops, conferences, professional training, and strengthening of research and potential trial sites. One of the major successes of this network has been the conduct of malaria vaccine trials at AMANET-strengthened sites in Africa.

The establishment of formalised networks that emphasise African collaboration such as ANDI and AMANET represent a new approach to addressing the research, innovation, and commercialisation challenges facing much of continental Africa. Their focus on building African health R&D networks to promote discovery, development, and delivery of locally suited technologies may be a critical stepping stone in encouraging stakeholders to work together in finding African solutions for African problems.

products and as a result we lack evidence for the notion that South Africa collaborates with other African countries to jointly develop health products based on their traditional knowledge.

8.5.3 Impact of entrepreneurial collaboration

The main impact we observed so far arising from South Africa's entrepreneurial collaborations is the distribution of its health biotechnology products in other African markets and its investment in the pharmaceutical sector. It is plausible that these products are more affordable than alternatives, particularly those from Northern countries, and also better aligned with local needs, but further research is needed. The heavy emphasis in the collaborations on marketing and the lack of almost any existing R&D collaborations show that South African collaborations in health biotechnology with other African countries have made almost no contribution to new-to-the-world innovation. There are no clear signs that this may change in the immediate future. However, market relations can be the first step towards closer R&D collaborations. The partners start to know each other and build up mutual trust. Networks can be formed and promoted by initiatives such as the SANBionet or ANDI, aimed at developing and promoting products based on African knowledge and biodiversity. Such initiatives may in the future contribute to new-to-the-world innovation as a result of African collaborations. Similarly, the presence of the ICGEB in South Africa may play a role in promoting entrepreneurial collaboration.

8.6 Conclusions

Our case-study research on South Africa's collaboration with sub-Saharan countries casts light on the main opportunities, challenges, and results of South-South collaboration in health biotechnology in Africa. Biotechnology is still nascent on the continent, and thus the levels of collaboration within Africa are low. However, our research also explores the potential role South-South collaboration can play on the continent. The main messages from this research are as follows.

South-South collaboration has strong political momentum in Africa. There are political forces promoting Africa-wide collaboration in science and technology, specifically in biotechnology. We see South-South collaboration on the agenda of the South African government, and more generally on the agenda of organisations such as the African Union and NEPAD. The focus on South-South collaborations in health biotechnology reflects a wish to see Africa progress in science-intensive fields, as well as the impetus to develop alternatives to North-based development models. As South Africa has the strongest capacity of all African nations in health biotechnology and is the continent's foremost publisher in the field, it is only natural that the political momentum for South Africa's collaboration with sub-Saharan countries includes a focus on health biotechnology.

South Africa's research collaboration in health biotechnology is focussed particularly on capacity-building efforts. With a growing number of sub-Saharan African countries targeting biotechnology as a field to prioritise, there is a growing demand for capacity-building efforts in the field. In general there appears to be a stronger emphasis on agricultural biotechnology rather than health biotechnology, but the latter is still a part of the package and can play a particularly powerful role in addressing African health problems. With South Africa nearby and its capacity in health biotechnology, the demand for capacity-building efforts is directed at South Africa. South Africa shares many health problems with neighbouring countries and its researchers understand the realities of science in the African context. The training South Africa provides is therefore likely to be both appropriate and cost-effective.

South Africa's entrepreneurial collaboration with sub-Saharan Africa is significant but almost entirely focussed on marketing ties. Our survey of entrepreneurial collaboration in health biotechnology showed South Africa to be in second place in terms of number of South-South firm collaborations of the six leading developing countries in health biotechnology. The collaborations are both with emerging economies, such as India, and with fellow African countries. Further examination has shown that South Africa's collaboration with the rest of Africa is heavily focussed on trade relations and almost never involves any development activities. So far, entrepreneurial collaboration is not a widely used tool for finding new solutions to local health problems, as there appears to be no contribution by the collaborations of new-to-the-world innovation. Still, with trade relations, bridges are built between African countries in health biotechnology, and with further efforts these could potentially be used to take the first steps in developing 'made in Africa' health biotechnology products and services.

To reap greater benefits of South-South health biotechnology collaboration amongst African countries, there is a need for support from external organisations. Despite recognising the importance of South-South collaboration in health biotechnology, South Africa simply does not have the resources to support collaboration significantly. Some of the strongest cases of South Africa's collaboration with sub-Saharan countries involve support from international organisations. As South Africa is in a position to support capacity building related to health problems in sub-Saharan Africa and is well attuned to local conditions, international organisations, philanthropic organisations, and Northern governments should harness these South-South collaborations in their development efforts and promote sub-Saharan Africa collaboration using Northern resources.

Our study shows that South Africa places an emphasis on South-South collaborations with fellow African countries and on promoting wider capacity in Africa in science-intensive fields. The level of collaboration in health biotechnology is still low, which is not altogether surprising considering how young the biotechnology sector is both in South Africa and in fellow African countries. As the sector matures in South Africa, there is increasing scope for the country to take on a significant capacity-building role. The continent has more than its share of health problems, and there is certainly demand for Africa-wide efforts to deal with them. The future is unknown and the key to the next vaccine or therapeutic may lie in the content's own ingenuity or biodiversity. There are therefore strong reasons to include health biotechnology as a part of the African renaissance, and to continue to promote South-South collaboration in order to invest in future African health solutions.

References

Al-Bader, S., S.E. Frew, I. Essajee, V.Y. Liu, A.S. Daar and P.A. Singer (2009). "Small but tenacious: South Africa's health biotech sector", *Nature Biotechnology* 27(5): 427-445.

Aspen (2009). *Aspen in East Africa*.

Biovac (2010). *Strategic international partnership*.

Boshoff, N. (2010). "South–South research collaboration of countries in the Southern African Development Community (SADC)", *Scientometrics* 84(2): 481-503.

Braude, W., P. Thandrayan and Sidiropoulos E. (2008). *Emerging donors in international development assistance: The South Africa case*. International Development Research Centre.

COMTRADE (2010). *United Nations Commodity Trade Statistics Database, commodity code H3-30*.

DIRC (2009). *Strategic plan 2010 – 2013*. Pretoria: Department of International Relations and Cooperation, Republic of South Africa. Available at *www.dfa.gov.za/department/strategic%20plan%202010-2013/strategic%20plan%202010-2013.pdf*

DST (2001). *A National Biotechnology Policy for South Africa*. Pretoria: Department of Science and Technology, Government of South Africa.

———. (2009). *Five point strategy*. Pretoria: Department of Science and Technology, Government of South Africa.

FAO (2010). *FAOSTA*.

Government of Zambia (2003). *Zambian national biotechnology and biosafety policy*. Lusaka: Government of Zambia. Available at *www.mstvt.gov.zm/index.php¿option=com_docman&task=doc_download&gid=13&Itemid=72*

Hanyona, S. (2003). "Zambia develops biotechnology strategy", *Environment News Service*. Lusaka, Zambia.

Jeenah, M. and A. Pouris (2008). "South African research in the context of Africa and globally", *South African Journal of Science* 104(9-10): 351-354.

Kilama, W.L., R. Chilengi and C.L. Wanga (2007). "Towards an African-driven malaria vaccine development program: History and activities of the African Malaria Network Trust (AMANET)", *American Journal of Tropical Medicine and Hygiene* 77(6): 282-288. Supplement.

Makinde, D., L. Mumba and A. Ambali (2009). "Status of biotechnology in Africa: Challenges and opportunities", *Asian Biotechnology and Development Review* 11(3): 1-10.

Mboya-Okeyo, T., R.G. Ridley and S. Nwaka (2009). "The African network for drugs and diagnostics innovation", *Lancet* 373(9674): 1507-1508.

Mokoena, R. (2007). "South-South co-operation: The case for IBSA", *South African Journal of International Affairs* 14(2): 125-145.

Motari, M., U. Quach, H. Thorsteinsdóttir, D.K. Martin, A.S. Daar and P.A. Singer (2004). "South Africa--blazing a trail for African biotechnology", *Nature Biotechnology* 22: DC37-41.

NRF (2010). *Binational agreements*. National Research Foundation.

Nwaka, S., T.B. Ilunga, J.S. Da Silva, E. Rial Verde, D. Hackley, R. De Vre, T. Mboya-Okeyo and R.G. Ridley (2010). "Developing ANDI: A novel approach to health product R&D in Africa", *PLoS Medicine* 7(6): e1000293.

OECD (2009). *Globalisations and the emerging economies: Brazil, Russia, India, Indonesia, China and South Africa*. Paris: Organisation for Economic Co-Operation and Development.

Ogodo, O. (2006). "Kenya approves a national policy on biotechnology", *SciDevNet*. Nairobi.

Pascal Newbourne Mwale, P.N. (2006). "Societal deliberation on genetically modified maize in southern Africa: the debateness and publicness of the Zambian national consultation on genetically modified maize food aid in 2002", *Public Understanding of Science* 15(1): 89-102.

Pouris, A. and A. Pouris (2009). "The state of science and technology in Africa (2000–2004): A scientometric assessment", *Scientometrics* 79(2): 298-309.

QS (2009). *2009 World University Rankings*. Qualiy & Success.

Republic of Kenya (2006). *A national biotechnology development policy*. Nairobi: Republic of Kenya.

Stevens, G., P. Chetty, J. Stout, T. Tarragona, T. Stiles, S. Molapo, C. Roodt, L. Khumalo, C. J and J. Gilmour (2009). "P06-07. A GCLP accredited Clinical Trial Laboratory Network in Africa and India: A collaborative effort between IAVI and in country research organizations", *Retrovirology* 6(3): 96. Supplement.

The World Bank (2010). *World development indicators database: Research and development expenditure (% of GDP)*.

Thorsteinsdóttir, H., A.S. Daar, P.A. Singer and E. Archambault (2006). "Health biotechnology publishing takes-off in developing countries", *International Journal of Biotechnology* 8(1): 23-42.

UNCTAD (2010) *Handbook of statistics 2009*. United Nations.

———. (2005). *Case study on outward foreign direct investments by South African enterprises.* Geneva: United Nations Conference on Trade and Development. Available at *http://www.unctad.org/en/docs/c3em26d2a5_en.pdf*

———. (1999). *Foreign direct investments in Africa: Performance and potential.* New York and Geneva: United Nations Conference on Trade and Development. Available at *http://www.unctad.org/en/docs/poiteiitm15.pdf*

UNDP (2010). *Human development report 2010 - 20th anniversary: The real wealth of nations—Pathways to human development*. New York: United Nations Development Programme.

9 Paving the Silk Road: Sub-Saharan Africa's South-South Collaboration with China and India in Health Biotechnology

AUTHORS: *Andrew Kapoor, Peter A. Singer, Joseph Wong, Halla Thorsteinsdóttir*

9.1 Introduction

The influence of China's and India's increasing global power is being felt across the African continent. Both countries have established linkages within many sectors in sub-Saharan Africa including: natural resources, infrastructure, information and communication, health care, and science and technology (Broadman, 2007; Goldstein *et al.,* 2006). Traditionally, much of Africa's history has been shaped by the influence of European nations and the outcome of the Berlin Conference, held in 1884–85, which laid out how Africa would be partitioned and formally colonised. Europe's 'scramble for Africa' was driven by the interest of wealthy European nations to dominate Africa's trade and control its resources (Fage and Tordoff, 2002; Zeleza, 1997). Colonisation came, but eventually so too did independence. In sub-Saharan Africa, it began with Ghana gaining its independence in 1957, and by 1966 most countries on the continent had achieved the same feat. After independence, the countries continued to have strong ties with many former colonial powers, and trade relations with Africa remain largely controlled by Europe (Zeleza, 1997; UNDP, 2009). However, half a century of failed aid and economic policies in Africa (Sahn *et al.*, 1999; Moyo, 2008), along with continued reliance on post-colonial ties has intensified the need for alternative solutions to the trade, economic, and human development problems facing Africa.

As both China and India are becoming global powerhouses, it is not surprising that they are playing an increasing role in Africa. The political and economic relationships they are forging with other countries in various areas are becoming particularly important to understand; China and India account for 37.5 per cent of the world population (The World Bank, 2010) and rank as the second and fourth largest global economies, respectively (UNCTAD, 2009). As a result, there is increasing interest in China's and India's growing collaboration with Africa.

China's and India's collaboration with Africa is not a contemporary phenomenon; Africa has economic ties with China and India that date back to the early days of the Silk Road. But it has been over the past decade that trade has increased and there has been a renewed emphasis on Sino-African and Indo-African relations. Africa's annual trade with India has soared from US $967 million in 1991 to US $35 billion in 2008; while over the same period, China's trade with Africa jumped from US $1.8 billion to US $106 billion, representing a 40- and 60-fold trade increase, respectively (Broadman, 2007; WTO, 2008; Schwab *et al.*, 2009; Toyoshima *et al.*, 2004). A noticeable reorientation of Africa's trade towards China and India has occurred over the past decade, and major African economies such as Kenya, Nigeria and South Africa list both China and India among their top five trade partners (Schwab *et al.*, 2009; Toyoshima *et al.*, 2004).

Evidence of strong collaborative ties can be seen across the private sector and significant Chinese and Indian investments have been made on the continent by firms who have invested billions in the manufacturing, telecommunications, and pharmaceutical sectors, such as Tata Steel (Mumbai, India), Essar group (Mumbai, India), ZTE (Shenzhen, China), and Holley-Cotec (Beijing, China). While entrepreneurial collaboration has traditionally focussed on trade in goods and investment, both China and India have extended their firm collaborations with sub-Saharan Africa into sectors including science and technology. There are now many established Indian and Chinese biopharmaceutical firms operating on the continent. As China and India have placed emphasis on their health biotechnology sectors, they are likely to be well positioned to move into

African markets in health biotechnology, share newly gained expertise, and build R&D capacity through collaboration with African countries.

Many nations in sub-Saharan Africa have also begun to develop science and technology infrastructure. As mentioned in Chapter 8, South Africa, for example, is nurturing a promising biotechnology sector with a considerable number of international ties (Al-Bader *et al.,* 2009; Louët, 2006) supported by both universities and public research institutions, such as the University of Cape Town (Cape Town, South Africa), and the Medical Research Council of South Africa (Cape Town, South Africa) (Motari *et al.,* 2004). Universities and research institutions within Kenya, Nigeria, and South Africa have become increasingly active in knowledge production and account for 41 per cent of Africa's total research output (Pouris and Pouris, 2009). Kenya, Nigeria and South Africa consistently rank as the three largest contributors to health-related research publications in sub-Saharan Africa, including in subfields such as 'Biology and Biochemistry,' 'Clinical Medicine,' 'Molecular Biology and Genetics,' and 'Immunology,' among others (Pouris and Pouris, 2009). However, even though within Africa these nations have shown research strengths, they have limited collaboration with each other, and only 5 per cent of peer-reviewed articles published from 2004–2008 involved institutions in more than one African country (Mboya-Okeyo *et al.,* 2009; Nwaka *et al.,* 2010). As discussed in Chapter 2, it is only in malaria research that we have been able to identify considerable regional linkages within Africa.

Most sub-Saharan African collaboration is still undertaken with the United States and the European Union, often with the leadership, funding, and ownership for the research residing outside Africa (Mboya-Okeyo *et al.,* 2009). To lessen dependence on Northern countries, Africans needs to find countries to work with that have similar research needs in health and biotechnology but also still have significant capacity in R&D to share. With foci on science and technology, and an international agenda aimed at strengthening African relations, emerging producers of biotechnology knowledge in the South such as China and India may play an increasingly important role in Africa's development.

To better understand the role of China and India's collaboration with sub-Saharan African countries in health biotechnology, we have examined the policy context and science and technology initiatives being taken by governments in sub-Saharan Africa to build ties with China and India. We also mapped the key research and entrepreneurial collaborations sub-Saharan African countries have with India and China in health biotechnology. Then, we focussed more precisely on specific collaboration initiatives, and examined the driving forces behind sub-Saharan Africa's collaborative efforts in health biotechnology with China and India, the challenges they have faced, and the impacts the collaborations have had so far. Our assessment focusses on three African countries, Kenya, Nigeria and South Africa, countries that have relatively strong science and technology capacity and are in different regions of the African continent (Box 9.1).

9.2 Governmental interest and support

China and India's collaborations with Africa have been controversial, and although a detailed discussion of these controversies lies outside the scope of our study, it is important to highlight them as they shape the broader context within which scientific collaboration of sub-Saharan Africa is occurring with China and India (McBride, 2008; Rotberg, 2008; Cheru and Obi, 2010; French, 2004). Much of the controversy has surrounded the topic of natural resources and resource exploitation. The past decade has seen an increase in criticism from the West about these partnerships, as widely publicised summits, high-level engagements, and frequent state visits have raised the profile of Chinese and Indian engagements in Africa. It should be noted that China's approach to Africa has been scrutinised more intensely than India's largely because of its no-strings-attached 'non-interference policy,' which imposes few political, human rights, or environmental conditions on African governments. This policy has lent itself to the criticism of Western governments who argue that this approach perpetuates corruption, poor governance, and human rights abuses. Overall, the perceived exploitation of African natural resources by China and India has motivated much of the negative discourse about their partnerships with Africa and given them the ironic title of being Africa's 'new colonialists'(McBride, 2008). This has not deterred China or India from continuing to pursue ties with African nations, but rather, has encouraged them to formalise collaborations through several high-level government agreements.

BOX 9.1

Case-study research on sub-Saharan Africa's collaborations with China and India: Methodology

To examine China and India's collaborations with sub-Saharan countries, we chose to focus on their collaborations with Kenya, Nigeria and South Africa. We focussed on six sets of bi-national collaborations where each bilateral collaboration was considered a single case. We examined cases of health biotechnology partnerships between: (1) India-Kenya, (2) India-Nigeria, (3) India-South Africa, (4) China-Kenya, (5) China-Nigeria, and (6) China-South Africa. We selected these countries because they are China and India's most active collaborators in the health biotechnology field; they differ greatly economically, socially and culturally; they represent three of the major economic communities of sub-Saharan Africa, namely the Economic Community of West African States (ECOWAS), East African Community (EAC), and Southern African Development Community (SADC); they are the largest regional economies in each sub-region; they represent varying stages of biotechnology policy development; and they all view the biotechnology sector as being of high importance, each having developed and passed a biotechnology policy. Studying nations from these three regional economic communities may provide insight into some of the key differences and potential roles that RECs play in supporting international science and technology cooperation with China and India.

We followed the same methods as in the other case studies discussed in this book and described in Chapter 1. We carried out scientometric examination of China's and India's co-publications with Kenya, Nigeria and South Africa; surveyed health biotechnology firms in China, India and South Africa about their collaborations with all low- and middle-income countries; carried out document analysis of policy documents and literature relevant to the topic of South-South collaboration; and interviewed 63 experts (Table 9.1) in Kenya, Nigeria and South Africa about their views and experiences of collaborating with China and India.

TABLE 9.1

Breakdown of number of interviewees in case-study countries

Countries	Number of interviewees
Kenya	18
Nigeria	22
South Africa	23
Total	63

China created the Forum on China-Africa Cooperation in 2000 (Beijing Declaration) and the China-Africa Business Council in 2005 in an effort to reinforce and accelerate existing and emerging partnerships. The biennial meetings of the cooperation forum are attended by delegates from over 45 African countries, heads of regional, international, non-governmental organisations, as well as influential entrepreneurs (Ministry of Foreign Affairs of the People's Republic of China; The Sharm El Sheikh Summit and 4th Ministerial Conference, 2009; Beijing Summit and the Third Ministerial Conference, 2006; Jia, 2006). As a result of the meetings in 2000, many African governments set up ministerial commissions to plan and coordinate the implementation of their evolving bilateral collaboration with China. In addition to the commodity-driven sectors, the forum focusses on other core areas such as cooperation in education, science and technology, and health. During the fourth meeting of the Forum in 2009, the China-Africa Science and Technology Partnership Program (CASTEP) was launched and will be administered through the Chinese Ministry of Science and Technology. The programme supports workshops and technological training courses, technology transfer, equipment donation, joint research projects, and the design of high-tech science parks. The partnership programme aims to foster 100 new joint research partnerships and provide 100 African post-doctoral scientists an opportunity to carry out research at Chinese institutions (Department of International Collaboration Ministry of Science and Technology, 2010). According to Chinese President Hu Jintao, 'China values its friendship with Africa. To strengthen unity and cooperation with Africa is a key principle guiding China's foreign policy.'

In addition to the CASTEP programme, China has signed more than 10 inter-governmental agreements on scientific and technological cooperation with African countries, including Kenya, Nigeria and South Africa. These collaborative agreements prioritise research projects in areas such as biotechnology, medicine, agriculture, energy, mining and manufacturing (Rotberg, 2008; Kaplinsky and Morris, 2009). Specifically within the area of health, China has committed to deepen its cooperation in malaria treatment and prevention. It will provide US $70 million worth of medical equipment and herb-based artemisinin anti-malarial drugs to

30 hospitals and 30 malaria clinics built by China across Africa (Jia, 2006; Naidu, 2007).

India's democratic ideals, shared colonial history, and involvement in the anti-apartheid movement of South Africa have also allowed it to forge strong ties with sub-Saharan Africa. When asked to define India's collaborative efforts with Africa, Prime Minister Dr. Manmohan Singh said, 'We don't seek to impose any pattern in Africa. It's for the African people to decide on their future.' India's strategy of economic diplomacy in Africa mirrors much of China's Africa policy; and like China, India actively collaborates with Africa's resource-based sectors (Goldstein *et al.*, 2006; Cheru and Obi, 2010). However, India has also promoted the establishment of South-South collaboration in science and technology with sub-Saharan Africa. Similarly to China, India has held a number of high-level government summits with Africa, focussing on issues shared by Asian and African counterparts in an attempt to provide structure and to focus its collaboration with Africa. For example, the India-Africa summit, held in 2008, was attended by leaders from 14 select countries chosen by the African Union. The meeting was intended to strategically strengthen collaborative ties in priority areas such as science, technology, R&D, capacity building, social development, and health (India-Africa Forum Summit, 2008a and 2008b). The output was the Africa-India framework for cooperation that covers issues of bilateral, regional, and international interest to both India and Africa, including their common positions on UN reforms, science and technology, intellectual property, trade and agriculture, among other areas.

The IBSA agreement between India-Brazil-South Africa has also fostered stronger ties between India and South Africa. As discussed earlier in this book, the IBSA agreement is a development initiative for promoting South-South cooperation and exchange which encourages linkages between the countries in many areas, including energy, education, health, trade, and science and technology (IBSA, 2005 and 2007; Mokoena, 2007; Puri, 2007). A South African policymaker highlighted the importance of such agreements stating that, 'we want to be a bit more proactive in the South and to drive our own agenda and look at things ourselves, take care of issues ourselves and not only to be reacting to the North's agenda or to be dependent on the North's agenda.' The IBSA agreement has provided a platform for dialogue and action on Southern issues and is a key policy in the promotion of South-South collaboration.

All three of the African countries focussed on in this chapter have prioritised the development of their biotechnology sector and as China and India have built up strength in biotechnology, it is therefore a possible sector for South-South collaborations. Among these countries, South Africa was the first to develop a policy on biotechnology. The biotechnology strategy for South Africa was written in 2001 and committed an initial US $69 million over four years (2004–2007) for biotechnology development. The strategy was set in place to catalyse the development of a 'bio-economy' to create commercially viable biotechnology products with local relevance (UNCTAD, 2004). One of the main results of this strategy was the creation of four Biotechnology Regional Innovation Centres (BRICs) in South Africa aimed at strategically cultivating the development of biotechnology clusters in the country (Al-Bader *et al.*, 2009; Louët, 2006). In 2009, South Africa's Ministry of Science and Technology established the Technology Innovation Agency (TIA) unifying the activities of the main technology-based enterprises in South Africa, including the BRICS, as well as the main biotechnology funders including the Innovation Fund (Government of the Republic of South Africa, 2007).

Following South Africa's lead, Nigeria formulated a biotechnology policy in 2005. Its mission stated that Nigeria shall 'as a matter of priority, initiate appropriate steps to explore the use of biotechnology for the benefit of Nigerians and thus ensure that Nigeria becomes one of the international leaders in biotechnology.' Nigeria has set up a national agency called the National Biotechnology Development Agency (NABDA) to implement the ambitious policy, and its main focus lies in health, agricultural, and industrial biotechnology development (Government of the Federal Republic of Nigeria, 2005). Kenya has also developed its own biotechnology policy in 2006, placing early focus on public education and awareness, international and regional collaboration, and setting up institutional and legal frameworks (Government of the Republic of Kenya, 2006). However, Kenya is still in the early stages of its policy implementation in comparison to South Africa and Nigeria and its support for biotechnology development is limited.

Although both China and India's involvement in Africa has been subject to criticism, it is clear that both governments are signalling their interest in collaboration outside commodity-based sectors. Both China and India have put science and technology consistently among their top priorities in bilateral agreements with countries in Africa, and the many summits, ongoing forums, and technology partnership programmes (Department of International Collaboration Ministry of Science and Technology, 2010) echo this emphasis (The Sharm El Sheikh Summit and 4th Ministerial Conference, 2009; India-Africa Forum Summit, 2008a and 2008b; South Center, 2006: 183-185). African governments continue to place emphasis on the economic growth and development opportunities in S&T intensive sectors such as biotechnology as priority areas in their collaboration with China and India (Government of the Republic of South Africa, 2007; Broadman, 2007; NEPAD, 2006; Kagame, 2006). Although none of the countries under study have specific policies on South-South collaboration in health biotechnology, the emergence of national biotechnology policies in South Africa, Nigeria and Kenya has created a policy environment supportive of such collaborations. This emergent policy environment will provide a platform for sub-Saharan Africa to engage China and India in health biotechnology collaboration.

9.3 The geography of sub-Saharan Africa's health biotechnology collaboration with China and India

9.3.1 Mapping research collaboration

To examine the levels and distribution of South-South collaboration in health biotechnology in sub-Saharan African countries, we mapped their research and entrepreneurial collaboration with China and India, and identified papers co-authored by researchers from China and India between 1996 and 2009 (see Chapter 1 for details of the methodology). We identified a total of 538 papers co-authored by researchers from one of the 47 nations in sub-Saharan Africa, and researchers in China or India from the SCOPUS database between 1996 and 2009. Both China and India had similar levels of collaboration with sub-Saharan Africa; the data showed that China co-authored 280 papers with sub-Saharan African countries while India co-authored 258 papers. The top five sub-Saharan Africa countries with the largest numbers of co-publications were the same for China and India. These were South Africa, Nigeria, Kenya, Ghana and Cameroon respectively (Figure 9.1). More than 50 per cent of these co-publications have occurred between 2005 and 2008, demonstrating the increase in collaboration between sub-Saharan Africa with China and India over the last five years.

FIGURE 9.1

The main countries China and India collaborate with in health biotechnology research in sub-Saharan Africa, based on the number of co-authored papers between 1996–2009

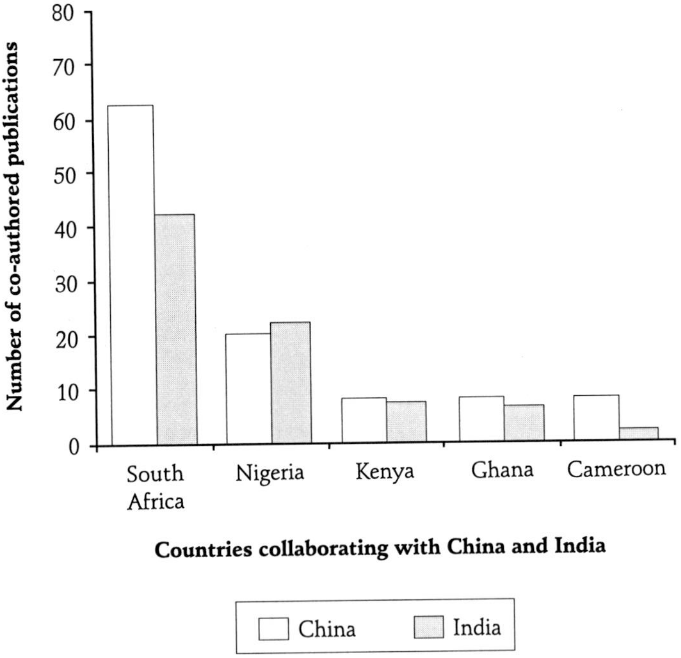

Source: Authors' presentation of data compiled by Science-Metrix using Scopus.

The data show that China and India have published more with South Africa than any other country in sub-Saharan Africa. Considering that South Africa is the strongest African country in health biotechnology and has the highest publication rate of all African countries (Boshoff, 2010; Jeenah and Pouris, 2008), this is not a surprising finding. South African researchers published a total of 104 joint research papers in health biotechnology with Indian and Chinese scientists between 1996 and 2009, Nigerian researchers published 42 joint papers, and Kenyan researchers published only 15 joint papers with India and China over the same time period. There does not seem to be a significant difference in the extent or distribution of China's *versus* India's co-publications with Africa, except China seems to publish slightly more health biotechnology papers with South Africa than India does.

The top three African countries differ somewhat in their more general South-South collaboration patterns. Nigeria's and South Africa's main Southern collaborators are China and India, in addition to Brazil in the case of South Africa, and South Africa in the case of Nigeria. But Kenya's main Southern collaborators are other African countries, South Africa, Tanzania and Uganda. Kenya's level of South-South collaboration in the health biotechnology sector is, however, so low that this pattern can easily change with increased collaborative efforts with another Southern country. Kenya places a relatively heavy emphasis on collaboration in health biotechnology with Northern countries with 78 per cent of its international collaboration with the North and 22 per cent of its collaboration with other Southern countries. South Africa does that as well with 84 per cent of its papers in collaboration with developed countries *versus* 16 per cent of its papers in South-South collaboration. In comparison Nigeria has 66 per cent of its collaboration with the North *versus* 34 per cent of its collaboration with Southern countries.

For all three countries, the most common subfield of health biotechnology for South-South collaboration was 'Genetics and Heredity.' This is the same subfield of health biotechnology that has the heaviest South-South collaboration in general as discussed in Chapter 2. The second most common subfield differed and was 'Virology' for South Africa, potentially reflecting the high level of the country's South-South collaboration in HIV/AIDS research discussed in Chapter 2. For Kenya the second most common subfield was 'Parasitology' and for Nigeria it was 'Biomedical Engineering'. 'Tropical Medicine' was a relatively common subfield for South-South collaboration for Nigeria and Kenya, in third and fourth place, respectively, but was only in seventh place for South Africa.

9.3.2 Mapping entrepreneurial collaboration

In order to examine the extent of entrepreneurial collaborations of sub-Saharan African countries with China and India we administered a survey to all Indian, Chinese, and South African firms that we could identify as being involved in health biotechnology activities and asked them about their collaborations with all low- and middle-income countries (see Chapter 1 for details of the methodology). We followed up the survey with inquiries and scoping in the field to capture any collaborations we may have missed in the original survey. We identified a total of 37 firm level linkages of the three focal countries of this study with China and India (Figure 9.2). The firm survey data showed eight entrepreneurial linkages with China, as compared to 29 firm linkages identified with India. The survey showed that the greatest number of collaborations occurred between South Africa and India, with a total of 12, followed closely by India-Nigeria linkages at 10. There were fewer linkages with China; South Africa, for example, only had four firm linkages one-third

FIGURE 9.2

The main sub-Saharan African countries China and India collaborate with in South-South entrepreneurial collaborations based on a survey of firms and scoping exercise

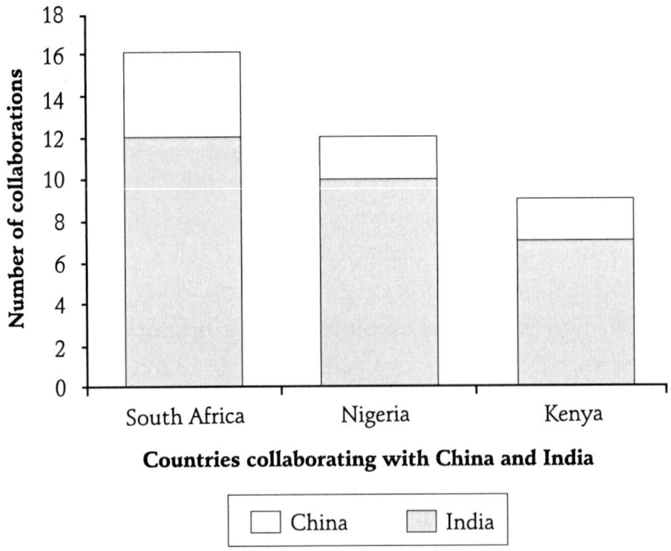

Source: Authors' presentation of data compiled by own survey.

the number of linkages seen with India. According to our survey, most of the collaborative activities with China and India were focussed almost exclusively on marketing, sales and distribution, with the exception of two South African firms who described joint R&D activities.

From this mapping we see two key observations. First, it is clear that both China and India are collaborating in research and entrepreneurial activities with the same countries in sub-Saharan Africa: Kenya, Nigeria, and South Africa. Second, when it comes to research collaboration, China's and India's levels of collaboration seems to be on par, whereas India is evidently participating more actively in entrepreneurial collaboration. However, it would be interesting to see how these linkages change over time.

9.4 Collaboration in research activities

To gain a better understanding of sub-Saharan African collaboration in health biotechnology, we interviewed researchers in Kenya, Nigeria and South Africa about their collaboration with China and India. All three countries have leading medical research institutes and universities with a degree of research output. In Kenya, the University of Nairobi (Nairobi, Kenya), as well as KEMRI (Nairobi, Kenya) are the country's leading institutions responsible for the production of health-related research (See Chapter 8). Nigeria also has a network of research-focussed universities (over 70) such as the University of Ibadan (Ibadan, Nigeria), the University of Lagos (Lagos, Nigeria), and Obafemi Awolowo University (Ile-Ife, Nigeria) in addition to national institutes such as the National Institute for Pharmaceutical Research and Development (NIPRD) (Abuja, Nigeria). NIPRD has successfully developed novel health products from traditional knowledge, such as Niprisan, a novel sickle cell anaemia drug (Wambebe, 2007). Both the institutions in Kenya and Nigeria have established collaborations with a number of leading institutions in India and China such as: the Council of Scientific and Industrial Research, CSIR (New Delhi, India), several universities, such as the University of Lucknow (Lucknow, India) and Zhejiang University (Hangzhou, China), as well as government ministries such as China's Ministry of Science and Technology (Beijing, China). Not surprisingly, South Africa has by far the largest and most internationally competitive network of biomedical research institutes, the Medical Research Council of South Africa (Cape Town, South Africa), Centre for the AIDS Programme of Research in South Africa (Durban, South Africa), the Council for Scientific and Industrial Research (Pretoria, South Africa), and leading universities like the University of Cape Town and the University of the Witwatersrand (Johannesburg, South Africa). These institutions have established scientific collaborations across the globe including strong links with China and India.

9.4.1 Reasons for research collaboration

As Africa is such a large and heterogeneous continent socially, culturally, and economically, we expected the motivations for engaging in collaboration with China and India would vary. Indeed, factors motivating the research collaborations we studied were diverse and varied by country. However, some of the more salient reasons for the research collaboration between sub-Saharan Africa and China and India that our interviewees stressed include:

1. *The need to test samples and access appropriate technologies:* Researchers in sub-Saharan Africa emphasised the need to fill gaps in their research created by an absence of specialised knowledge and equipment by sending samples to researchers in China and India. For example, they expressed the need to send samples for chemical testing to China and India. These included samples such as plant isolates, synthetic compounds they had developed, or other biological agents that needed further analysis. Several researchers at universities and public institutions in South Africa, Nigeria and Kenya had established these types of collaborations with scientists at the CSIR in India, citing them as critical to the advancement of their work. South African researchers have also begun to collaborate with China, sending compounds they have developed to be tested specifically in animal models. The main reason for the need to send samples to China and India was the dearth of high-technology scientific equipment in sub-Saharan Africa, particularly in Kenya and Nigeria. Many expressed a need to gain access to equipment such as nuclear magnetic resonance (NMR) and high performance liquid chromatography (HPLC) machines

that are critical in the identification of basic compounds, synthetic chemistry, and drug development. One Nigerian researcher emphasised this motivation for sending samples abroad and said,

> my samples are being tested in an Indian lab for diabetes, it's much cheaper than conducting the same experiment here...there are so many more institutions (in India) that do animal work it's much easier (there) because here we have to write many applications and it's very expensive.

Although the nature of the technology and sample exchanges varied by the sub-Saharan African countries; researchers viewed both China and India as equally suitable collaborators.

2. *Access to training and capacity-building opportunities:* Another critical motivator for researchers in sub-Saharan Africa to pursue collaboration was the opportunity for research personnel from Africa to visit institutions in China and India for training and capacity building. Researchers said that the ability to travel to China and India and to gain exposure to new information, research strategies, and complementary expertise, to pool resources, and to seek scientific advice was necessary to the advancement of their research projects. They noted that participating in such exchanges with China and India provided a familiar scientific landscape which helped translate directly into a rich educational experience. Scientists from sub-Saharan Africa found counterparts in China and India working on projects more closely aligned with their work in communicable diseases (HIV, TB, malaria) as well as areas of traditional herbal medicines. Researchers felt these types of expertise were more relevant than the expertise they gained through collaboration with Northern countries: as many of the techniques and strategies had been developed in a resource-constrained environment. One researcher commented that,

> It appears in the western world the focus is different. I would rather, in fact, send a student of mine to India to [a] lab to go and learn some basic things or to evaluate some similar things rather than send student to the US where perhaps we may be rather talking at some very high molecular level, which, I believe, is not the thing that is needed in this environment.

Many researchers emphasised that working with China and India, where scientists understood the challenges of conducting research in the developing world but had also amassed a wealth of expertise and knowledge in health biotechnology, was an important motivation for pursuing South-South collaboration.

Interviewees also highlighted the important role the Chinese government has played in providing educational opportunities to young African students, particularly from the less-developed African states. The Chinese Scholarship Council offers scholarships to both Chinese citizens wishing to study abroad and to foreign citizens wishing to study in China, with a particular focus on educational, scientific, and technological exchanges. In 2007 it provided 2,733 scholarships to African students, comprising almost one-third (27 per cent) of the total number of scholarships granted worldwide, the vast majority of which are grated in science, engineering and medicine. Africa received the second largest number of scholarships globally, second only to Asia itself (Hassan, 2007: 433-436; Government of the People's Republic of China, 2007). Many South African universities such as the University of the Witswatersrand, University of Pretoria (Pretoria, South Africa), and the University of Cape Town have received Chinese students studying on the Chinese Government scholarships. This has created a bi-directional exchange whereby Chinese students are also benefiting from programmes at leading African institutions. Our interviewees said that a high proportion of African students return to Africa after furthering their education in these countries as compared to those who travel to the west. One university researcher stated, 'The future is very bright because if you see the number of students that are going to China and India, that would give you an idea that the future is bright for Africa'; emphasising that this may begin playing a larger capacity-building role in the African research context.

According to our interviewees, organisations such as TWAS have also been key promoters of research exchanges. They have dedicated funds to facilitate both scientific and educational exchanges, which have benefited many scientists, particularly in

Nigeria and Kenya. These programmes have helped scientists across sub-Saharan Africa advance their work in Chinese and Indian labs in acquiring new skills and techniques, building capacity, and gaining international experience. As discussed in the chapter on South Africa, the ICGEB has begun to play a similar role in South Africa since setting up a branch in Cape Town to complement its New Delhi, India and Trieste, Italy offices. When comparing China and India, it appears that China has been more active than India in providing educational opportunities; whereas India seems to be more focussed on providing opportunities for visiting research fellows and professional exchanges. Now India is increasing its role in supporting education in Africa with 146 Raman fellowships aimed at supporting East African post-doctoral and research fellows in India. However, in both cases it is clear that third-party organisations such as TWAS and ICGEB have played a key role in brokering collaboration.

3. *Advancement and exchange in traditional herbal medicine:* Several researchers in sub-Saharan Africa expressed a keen interest in the development of traditional herbal medicines, and we identified several scientists carrying out collaborative work in this area. Researchers in all three countries cited the economic benefits that could be realised through the licensing and commercialisation of compounds screened for drug development in India or China. However, interviewees also emphasised that although Africa has a strong culture of traditional medicine, it remains informal and lacks the scientific grounding needed to exploit its potential value. Researchers in Africa identified India particularly as a key partner in helping African scientists build the scientific base for a traditional medicines industry. Scientists underscored that collaboration with Indian scientists and institutions and to a lesser extent the Chinese, has built capacity and advanced African knowledge and expertise in this area.

Our research highlighted specific areas of collaboration in traditional herbal medicine which can be grouped broadly into two areas: isolation and screening of active constituents in plant extracts; and analysis or synthesis of compounds (see Box 9.2).

Interestingly, South African researchers engaged in a different type of collaboration with respect to herbal medicine. They preferred joint work where they could learn from Indian counterparts about regulation, patent protection issues, standardisation of traditional medicinal products, management of traditional knowledge databases, distribution of benefits, and the prevention of exploitive science. India has made strides in helping South Africa set up an electronic database to manage traditional knowledge comparable to the one set up by the CSIR in India (Dickson, 2003).

BOX 9.2

South-South collaboration in traditional medicine: (Nigeria–India)

The National Institute for Pharmaceutical Research and Development (NIPRD) has maintained longstanding collaborations with herbal scientists at India's Council of Scientific and Industrial Research (CSIR), and has carried out a number of collaborative projects exploring the efficacy of compounds identified in Nigerian flora with specialists in India. Their research collaborations in traditional herbal medicine have focussed both on isolation and screening of active constituents in plant extracts, and on the analysis and synthesis of compounds. Elsewhere in Nigeria a group of University researchers were also interested in acquiring traditional knowledge through technology transfer from firms in India. An Indian firm was working with researchers from the University of Ahmadu Bello, Zaria, Nigeria to transfer knowledge and technology related to the Neem plant. Researchers were interested in learning how to take advantage of the locally grown plant that has been used in Indian Ayurvedic medicine for centuries, and were working on transferring extractive technology, purification protocols, product formulations, and packaging. The Nigerian researchers were looking to harness the Neem oil properties to produce various anti-viral and anti-fungal products locally.

4. *Common health concerns:* Researchers in sub-Saharan Africa identified collaboration with China and India as a mechanism to increase research on local problems that have low priority in the North and subsequently lack funding and political will. They emphasised a strong common research focus on disease areas like HIV, TB and malaria, the major diseases which affect Africa, China and India.

One researcher emphasised the importance of setting research priorities with a Southern focus saying that, 'if you do research that others are not interested [in] you will not get anything. I think that if malaria was a problem of the west, there would have been a vaccine in market now, but because it's not their problem we have to engineer that research in the South and solve the problem ourselves.' Researchers in our study cite South-South collaboration as a main strategic advantage in the path to finding appropriate, low-cost solutions that work in the challenging conditions found in many African countries. While discussing the importance of such a Southern advantage, a Kenyan scientist also pointed out that, 'You normally go where you are going to be getting the solutions. So the people who understand the problems the best are going to offer you the best solutions.' The IBSA initiative has also encouraged a research focus on common health concerns. Scientists in India, Brazil and South Africa have joined forces as a part of the IBSA initiative to tackle some of these key disease areas with teams working on ways to reduce HIV infection using novel treatments, develop TB diagnostic tools, and a malaria vaccine (Mokoena, 2007).

5. *Lessen dependence on the North:* Researchers in sub-Saharan Africa also emphasise the need to lessen their dependence on the North as a motivation for pursuing South-South collaboration. There is strong interest among researchers we interviewed to shift away from a North-based model in favour of a more Southern approach. This view was particularly emphasised by South African interviewees. South Africa has been successful in acquiring research funds through the highly competitive European framework programme securing 21 projects with FP6 (European Commission, 2010). However, the framework's primary goals are 'to improve European science' and encourage third country participation only as a distant secondary objective. Researchers and policymakers suggest this as a risk for developing countries that may be easily swayed from local, regional, and national priorities in favour of a Northern-focussed research agenda. It was also stressed that the role of African researchers cannot be confined to being providers of

research material, and as a South African policymaker stated '...if you want to test our people we want our researchers to be part of it. Not only as guinea pigs but to be involved right back from the basic sciences.' The strong research base in South Africa has allowed researchers there to be more selective in collaboration and thus less likely to accept collaborative arrangements that do not align with their scientific priorities. Researchers note that China and India provide attractive alternatives to the North and are able to provide collaboration on more relevant and equitable terms.

In contrast, researchers in Kenya and Nigeria find it difficult to lessen their dependence on the North as their access to funding is limited. Researchers feel it is harder to negotiate terms and a local research agenda when they have minimal funds to contribute towards joint research projects. Consequently researchers in these countries state they are more likely to be involved in research collaborations that limit their contributions to carrying out data collection tasks rather than being involved in protocol development and experimentation. Researchers in Kenya and Nigeria further expressed concern over the marginalisation of their work due to the poor international scientific reputations of their countries. They emphasised their difficulty in getting their publications accepted by international journals and noted that this challenge pushed them to seek collaborators to increase the acceptance of their work. Where traditionally these may have been researchers from former colonial powers they include Chinese and Indian scientists with increasing regularity.

9.4.2 Challenges

Below we discuss the main challenges for collaboration with China and India, as described by our interviewees.

1. *Lack of funding:* Financial support for scientific research and joint scientific collaborations comes from a variety of sources but is quite limited in most African countries. None of the countries we studied had funds dedicated to support collaboration in health biotechnology research. Governmental funding for basic research

is extremely limited in Africa, and funds for collaboration must often be squeezed from other areas, making it difficult for African researchers to initiate collaborative projects. When speaking about funding collaborative work one researcher stated,

> it comes from my salary you have to save towards it...so for example I have students who must travel by the end of this year, I have started saving for them, I must because I believe in the work that they are doing and I believe that this kind of collaborative work is going to benefit us.

Researchers said that the lack of funds prevents collaborative work from continuing after a research fellowship as the resources necessary to continue the joint work are not available in their home countries. Several researchers who participated in research exchanges were disappointed that their collaborations were hampered when they returned back to Africa because no funding was available to continue. One researcher stated, 'We are trained, we gained the skills but we are not able continue our work. In China I get the state of the art equipment, I come back here and I don't have the same infrastructure or funding, so how am I going to continue to do that research?'. As discussed in Chapter 8, the ICGEB has begun to change this model by providing funding for scientific exchange and further providing seed grants for researchers returning home, allowing them to continue their work.

Even though South Africa fares significantly better than the rest of Africa in terms of funding for science and technology, with a larger national budget and dedicated funding for biotechnology, researchers noted that they still face several challenges in funding collaborative research. Although funds are more easily available, they are broken up into numerous small grants that prevent collaboration from moving beyond the basic science level, which is available in South Africa, to the proof of concept stage. This makes it necessary to seek funding from larger pots that are often being offered by developed countries through, for example, the European Union framework programmes or the US National Institutes of Health (NIH), encouraging them to collaborate with well-established Northern partners rather than others in the South. Researchers stated that larger grants dedicated to collaborative research are needed to help develop larger scale, higher-level collaborations that they currently cannot establish easily with countries like China and India.

Researchers also highlighted that some of these barriers could be overcome through third party organisations such as the TWAS and ICGEB, identifying them as critical players in the provision of funds specifically for research collaboration. TWAS, ICGEB, and the European framework programme have been strong funders of basic scientific collaboration, providing research scientists with funds to carry out joint work across the developing world. Researchers explained that collaborations with emerging economies such as China and India still face major funding challenges as collaborators in those countries often cannot afford to fully fund joint projects as many Northern partners are able and willing to do. As one researcher stated, 'In the South people are much poorer than those we are talking about in the North, so when it comes to making funds available, they are not as readily available as they would be from our traditional Northern partners.'

2. *Scientific and supporting infrastructure:* Infrastructure still remains a key barrier to both research and research collaboration in sub-Saharan Africa. With a shortage of equipment, reagents, and laboratory facilities many scientists find it hard to pursue the type of research they have been trained to do. One scientist said that, 'Personally I started with organic synthetic chemistry and ran into problems with analytical equipment and so it was easier to go into the natural products industry because it is not equipment intensive.' Even when equipment is obtained through grants and assistance from Northern counterparts, scientists find it hard to maintain the equipment. It often requires carefully controlled temperature and humidity conditions or a constant power supply, which can be difficult to ensure, especially in countries that suffer lengthy power outages such as Nigeria. Researchers indicate that these local barriers thwart collaboration as it disables otherwise productive researchers and either prevents them from carrying

out their portion of collaborative work or dissuades collaborators from engaging them. Researchers further state that although they collaborate to gain access to infrastructure, there is a base level of local equipment necessary for them to be able to engage in any research. Without this basic level of infrastructure in place they are not able to add value to joint research projects, which makes it difficult for Chinese and Indian researchers to justify the benefits of collaborations with African scientists.

Conversely, South African researchers with higher levels of basic research infrastructure cite they are often left frustrated at international conferences and meetings by being 'lumped with the rest of Africa'. They cited only highly specific technical equipment such as proton NMR as a challenge to South African research which necessitated collaboration. South African researchers emphasised that their being labelled as 'African' undermined their research capacity and thus tended to generate scepticism about their competencies when courting potential collaborators. In contrast to researchers in Kenya and Nigeria, some researchers in South Africa highlighted that they felt China and India were less developed then South Africa in terms of scientific infrastructure. This interesting reversal further illustrates the contrast between South Africa and other African countries.

The cumulative frustrations, including working in a disabling environment, drives many bright young African researchers to find posts in Europe, the United States, or Canada where they may pursue collaborative work or further education. One researcher stated, 'you would find that in Africa we train a lot of people, there are a lot of good researchers. But they are no longer living in Africa.' Subsequently many decide to stay abroad, leaving large gaps at even the top research institutes in sub-Saharan Africa. This effectively limits the number of prominent scientists who are available to collaborate between top African institutions and those elsewhere in the South. South Africa has particularly suffered the impact of this phenomenon and though it is ahead of the rest of Africa in terms of its facilities, funding, and resources, it lacks the equivalent force of skilled scientists and researchers. The absence of these eminent researchers who work for long periods of time abroad has ultimately impacted Africa's ability to initiate, sustain, and leverage high level scientific collaboration.

9.4.3 Impacts of research collaboration

When we asked researchers in Kenya, Nigeria and South Africa to indicate the main impacts of the collaboration, a common response was that it led to: (1) capacity building adjusted to their needs and conditions, and (2) the ability to focus on a local research agenda. Respondents stated that these two key areas increased publications, allowed the scientists to focus on a locally relevant research agenda, provided the opportunity to acquire new knowledge, and allowed for capacity building. Respondents stated that these were key areas where South-North collaboration comes up short, and two specific areas where South-South collaboration may have huge advantages. A researcher commented on the impacts of their collaboration with scientists in India stating that:

> The ability to learn new skills, the ability to hone the skills that you have already and then the ability to—if I may use the phrase—better yourself. When you come back you are able to do more than you were able to do before and have the ability to publish papers on whatever you have done. I think for me these have been big successes of my collaboration.

This highlights that there are important benefits sub-Saharan Africa countries stand to gain from the Chinese and Indian biotechnology. Scientists cite that the collaboration has made it possible for them to be able to publish in higher impact journals and thereby increase the visibility of their research. Another common impact was the ability for researchers to access training opportunities in the form of fellowships, research sabbaticals, conferences, and additional degrees for both themselves as well as their graduate students. Researchers felt this collaboration was slowly building skills and knowledge within the various research communities across Africa. Researchers also cite that there has been a significant increase in collaborations between sub-Saharan Africa and China and India in health biotechnology specifically over the past few years, which is supported by our analysis of joint publication between these regions. Unfortunately, building capacity and developing a stronger local knowledge base takes years to materialise tangible results. Measuring the impacts of research collaborations is similarly difficult as there are few tools that are able to measure the complex intangible benefits that accrue during the course of such collaborations.

9.5 Collaboration in entrepreneurial activities

Mapping of entrepreneurial collaboration between sub-Saharan African countries and China and India showed firms involved varied widely in their business structure and included locally-owned firms and Chinese and Indian-owned subsidiaries. Also, as mentioned above our research illuminated two consistent trends at the firm level. Firstly, there were far more Indian firms collaborating in health biotechnology with sub-Saharan African countries than Chinese firms. Secondly, the data showed that collaborations between India, China and sub-Saharan Africa are heavily focussed on two main activities: marketing and distribution and the provision of supplies (active pharmaceutical ingredients, formulations, packaging, equipment etc.). There was little evidence of joint innovative activity and the marketing, distribution, and supplies seem to be moving largely from China and India towards Africa. South Africa appeared to be the most common partner of both China and India followed by Nigeria and Kenya.

9.5.1 Reasons for entrepreneurial collaboration

Interviewees indicated several key factors that motivated their collaboration at the firm level.

1. *Access to markets and products:* Both the survey and the follow-up interviews showed market access to be the major motivation for both Indian and Chinese firms to collaborate with Kenya, Nigeria and South Africa. This result is consistent with China and India's recent efforts to bolster trade with African markets (WTO, 2008; OECD, 2006). Interviewees in these three African countries also indicated that access to their markets provided gateways more broadly into Africa for the distribution and sale of health and pharmaceutical technologies from China and India. This was particularly important for Indian and Chinese firms in Nigeria

BOX 9.3

Advancing South African biotechnology through collaboration: (South Africa – China)

South African firms are involved in more advanced types of collaboration in comparison to many of their counterparts in sub-Saharan Africa. They encompass a wide range of activities outside simple marketing and distribution including: manufacturing, vaccine research, bioinformatics, and gathering data for clinical trials. The South African firm Altis Biologics (Pty) Ltd. (Pretoria, South Africa) works for example with the First Affiliated Hospital of Xinjiang Medical University (Xinjiang, China) on bone morphogenetic proteins and their effects on bone induction and bone remodelling in rabbits to gather pre-clinical data. Altis Biologics specialises in the R&D of osteogenic biomaterials for use in skeletal regeneration therapies and has developed a new osteogenic biomaterial for use in healing fractured bone. However, owing to expense and lack of local expertise, Altis forged collaborations with Xinjiang Medical University. It supplied their newly developed bone regeneration material to Chinese researchers with expertise in orthopedics and in carrying out animal trials to advance their understanding of its capabilities *in vivo*. This helped the South African firms gather valuable pre-clinical data and meet the requirements of South Africa's Innovation Fund (now part of Technology Innovation Agency) to apply for funding to move the product to clinical trials.

and South Africa, two of Africa's largest markets. In addition, South African firms also viewed China and India as significant markets for the distribution and sale of locally developed South African technologies and products. It is worth noting that several well-established firms are also preparing to move from marketing and distribution into local production including Dr. Reddy's Laboratories (Hyderabad, India) and Holley-Cotec (Beijing, China).

Our research underscored that while Indian firms did not seem to be involved in any specific type of product, Chinese firms seemed to focus on the niche market for malaria therapeutics. Sixty per cent of the Chinese firms interviewed in our study are involved in the marketing, distribution, or production (through sourcing of active pharmaceutical ingredients) of artemisinin-based products for treatment of malaria. Entrepreneurs cited this as a strategic advantage for firms collaborating with China, as they have significant strengths in this area. Chinese entrepreneurs in Kenya also stated that this niche area developed when China, a leader in artemisinin-based products, no longer had a domestic market for malaria medications. A Chinese entrepreneur in Kenya stated, 'we share our Chinese experience on malaria control and prevention, 30 years ago China also had a malaria epidemic. But right now there are no more malaria cases in China; I think this is a very good experience to share with Africa.'

2. *Access to technologies and technical expertise:* Despite the low numbers of entrepreneurial R&D collaborations, African firms cited the importance of collaboration in gaining access to the knowledge and technology that China and India have to offer (Box 9.3). We found India to be an important supplier of technologies to sub-Saharan Africa, including technologies for low cost generic drug production, specifically for diseases such as TB and HIV, as well as a key supplier of traditional herbal medicines. Both China and India have begun to establish a record of innovation in health biotechnology and an increased ability to share their experiences of innovation (Frew *et al.*, 2008 and 2007). For example, East Coast Rapid Diagnostics and Life Assay (Durban, South Africa) are joint ventures between the publicly funded LIFElabs (Durban, South Africa) and the Indian Tulip Group Diagnostics (Bambolim, India). The Indian company agreed to transfer several diagnostic technologies to South Africa including rapid malaria diagnostic kits, pregnancy diagnostic kits, and urine dip stick technology. A formal agreement was signed where Tulip Group Diagnostics committed to transfer not only the technology but also provide substantial capacity and technical assistance. The Tulip Group assisted with the establishment of a local South African manufacturing operation which will produce the kits for use in South Africa as well as other African nations with high rates of malaria and infectious disease. Officials from the Ministry of

Science and Technology in Nigeria reinforced the idea that, 'The main reason for collaborating is to acquire the technology. We feel it is much easier and cheaper to get biotechnology from the South than from the North and we feel we can quickly learn and be at par if we deal with the South.' However, not all the benefits lie in Africa. Chinese and Indian firms benefit in several key ways by opening up new markets; they are able to expand their customer base via increased market access and revenues.

3. *Cost advantage of doing business:* Most biotechnology firms in Africa are not engaged in intensive R&D activities and thus not developing new-to-the-world innovative products. Though there is some collaboration in manufacturing, it is mainly in the filling and finishing of products. Many firms cited that the high domestic costs of manufacturing in Africa have limited the scope of such collaborations and favoured manufacturing sites remaining in China and India. Further, the availability of low-cost technologies and generics from China and India, thus, heavily favours collaborations in marketing and distribution as a more cost-effective strategy for firms in sub-Saharan Africa at this stage.

The firms in our study emphasised that to stay competitive in such markets, they needed to keep costs down; and interviewees emphasised that the shift in their business collaborations towards China and India was driven by a comparative cost advantage relative to developed countries. One entrepreneur in Nigeria who sources vaccines from Shanta Biotechnics (Hyderabad, India) in India stated:

> If we collaborate with Southern partners it makes a lot of sense for us because without Shantha vaccines if we had to buy the hepatitis vaccine from GSK for example, you can imagine what people will pay. I think that South-South model that we have in place with India is going to help us to crash the prices and make these products more affordable for all those that need them.

At present, many firms find it preferable to access products and technologies through China and India as their main method of generating revenue, and the strong overall message was that China and India are able to provide African countries with the requisite products at prices far lower than developed countries are able to.

When firms were asked to compare their collaborations with China and India, many cited a main advantage in collaborating with India was its ability to offer goods on credit. This was echoed by capital-poor firms, particularly in Kenya and Nigeria as a huge advantage over collaborating with China, which provides fewer credit options. One firm commented that, 'we look for any sort of collaboration whether it be R&D or it be technology transfer, obviously looking to keep costs down so that we can satisfy our mission of providing cost effective, affordable therapeutics. And that's where I see particularly India becoming very very important in the future.'

9.5.2 Challenges for entrepreneurial collaboration

1. *Scientific and supporting infrastructure:* While lack of scientific infrastructure pushed researchers into South-South collaboration, the lack of more basic infrastructure limited the opportunities of firms in engaging in South-South collaboration. Most countries in sub-Saharan Africa face huge challenges without the basic infrastructure necessary to support commercial activity in health biotechnology. The firms cited poor roads, fuel shortages, sporadic electricity, reliance on generators, and difficulties in clearing goods as the main challenges in this respect. The main impact felt by firms was an overall increase in the end cost of their products and challenges in exchanging goods with their collaborators. In Nigeria, power is a particular problem for companies, especially those with voltage sensitive scientific equipment 'Electricity is not reliable here it's zero almost all the time. Right now we're running on the generator. So out of 24 hours I think we hardly get four hours on average.' Firms in South Africa, Kenya, and Nigeria also identified secondary factors such as: inefficiencies in transportation, communication, and travel, as areas that slow down and discourage collaboration between sub-Saharan Africa, China and India. They noted that this makes collaboration increasingly

difficult as operations, communication, and manufacturing can become unpredictable and therefore unattractive to potential collaborators from China and India.

In contrast to firms in Nigeria and Kenya, South African firms cited China and India's poor infrastructure as a challenge to collaboration. South African firms cited concerns with infrastructure in China and India as an impediment to maintaining high standards of quality and practice, thus making potential South African collaborators skeptical of collaborating with Chinese and Indian firms. This highlights the substantial differences between South Africa's stage of infrastructure capacity and that of other countries in sub-Saharan Africa (Motari *et al.*, 2004).

2. *Lack of understanding of the local markets:* Indian and Chinese firms that set up collaboration with partners in sub-Saharan Africa often have little experience of the realities and costs of operating in these countries. Chinese nationals we interviewed in Kenya highlighted that many Chinese business attempts fail in Kenya because entrepreneurs do not understand the African markets before they arrive. When asked about the characteristics of successful entrepreneurs, strong English skills and flexibility were identified as being vital. Some Indian nationals we interviewed in Kenya emphasised similar challenges facing Indian entrepreneurs, stating, 'the problem is one of perceptions and lack of information' among Indians wanting to collaborate with firms in Africa. Interestingly, many entrepreneurs in sub-Saharan Africa saw India as more receptive to the needs of the African markets in comparison with China. One entrepreneur noted that, 'India has been flexible. Say with packaging—African markets need market specific packaging. They are able to develop small volume products to maintain African regulatory requirements which companies elsewhere find hard to do.'

   Understanding specific areas such as marketing has proved challenging for many firms, as they do not understand how to reach consumers in these countries, and thus, fail before they are able to gain any market share. One interviewee stated,

> There are many examples of Indian companies that have tried to open up in South Africa and have failed, or have taken decades and decades to achieve a small market share. That is because of the differences in markets, somebody coming out of India to manage a business in South Africa doesn't understand the market here as well as South African businesses do.

Firms in our study interested in Chinese and Indian markets also cited difficulties, especially in adapting to the intensely competitive business environments in those countries.

Language barriers were stressed as a factor limiting the understanding of African markets and posing special challenges to the collaboration. Many firms felt collaborations with China were hindered by the language barrier they faced in dealing with Chinese firms and institutions. Some firms cited that these barriers lead to misunderstanding of each others' capabilities, resulting in misalignments within the collaboration. However, interviewees and firms noted that their Chinese collaborators were working hard to minimise the negative impacts and prioritising training in English.

In comparison, firms in sub-Saharan Africa cited language and culture as an advantage to collaborating with India. One interviewee stated, 'With India perhaps, we are better because we speak the same language, Nigeria and India were both colonized by the English, so we speak English. With China we usually have problems.' Additionally, the large diaspora communities of Indian ancestry in many parts of Africa were also cited as reasons that firms felt more confident in their understanding of Indian commerce and culture. Many firms in our focal countries had employees of Indian decent and felt they 'knew Indians well' and further expressed that they could relate both to their culture and style of doing business because they had personal experience with them.

3. *Difficulty in identifying reliable collaborators:* Though interviewees from firms and key informants agree that South-South collaboration is of further importance, they cite identifying

partners as a major challenge. One of the main barriers remains that few firms in Africa know what is happening in India and China. They have little opportunity to gain international exposure and find it difficult to keep tabs on the rapidly changing biotechnology landscapes of both China and India. One firm's CEO stated, 'I know there's language barrier but we don't know what they are doing, we don't really know what's going on in China.' This presents firms with an immense challenge when looking for partners with complementarities that will aid in developing compounds, technologies, services, or ideas into commercial goods.

Firms also cited trust as a stumbling block in identifying partners. Most firms in sub-Saharan Africa do not have the capacity to conduct due diligence on potential collaborators, and government bodies, particularly in Kenya and Nigeria, are rarely able to assist them in the process. Performing due diligence can be very expensive and though many firms in South Africa described carrying out such a process when identifying collaborators, those in Kenya and Nigeria find doing this financially impractical.

The risk of encountering counterfeit medicine in African markets, for example, drugs that lack active ingredient or have passed their sell-by date, underscores the importance of identifying reliable collaborators from China and India. Many governments in sub-Saharan Africa lack the regulatory capacity to verify the safety and efficacy of drugs and make sure that substandard drugs do not reach their markets. Consequently some African countries have become prominent recipients of substandard products from both China and India. The main challenge firms identified was having counterfeits of their own products undercut them in the marketplace. These counterfeits are sold at lower prices and can subsequently attract huge market shares. The threat of counterfeit products in African markets has also made Indian and Chinese collaborators reluctant to enter into marketing and distribution agreements with firms in sub-Saharan Africa.

4. *Corruption and bureaucracy:* Firms in sub-Saharan Africa cited that conducting business in an environment where corruption and

excessive bureaucracy are common problems poses a huge barrier to collaboration. Respondents stated, 'political favouritism and nepotism often determined business opportunities to collaborate with companies in China and India.' Many firms listed an important aspect of this practice being visible in the government tender system and in purchasing agreements arranged by sub-Saharan Africa governments. South African firms also cited the extremely long processes and bureaucracy involved in forging collaborative projects and arrangements, especially with India. A South African policymaker reinforced this idea stating, 'It is extremely difficult to get permission to do anything. It can actually kill the projects. By the time we get permission to do something you have lost the will to do it. That is one of the major problems we've got [with India].'

9.5.3 Impact of entrepreneurial collaboration

Interviewees cited that collaboration with China and India in the field of health biotechnology provides a good mechanism to increase the availability of vaccines, pharmaceuticals, and diagnostics to the wider population in sub-Saharan Africa. The impacts of such collaborations have driven down the cost of basic essential medicines and opened up African markets to technology that has historically been too expensive for widespread deployment in low-resource settings. Examples include the marked drop in the cost of hepatitis, TB, malaria, and HIV treatments in Africa available through generics producers in China and India such as Cipla (Mumbai, India), Ranbaxy, and Holley-Cotec (Beijing, China). Cheap diagnostics, vaccines, and generics jointly developed or simply acquired through collaboration with China and India could thus continue to positively improve access across Africa.

According to respondents, traditional herbal medicine may hold potential for African firms. Although not many products have yet made the leap in Africa from research to marketplace, our interviewees' evaluations indicate that we may expect more impacts in this area in the future. Ongoing research into plant extracts and various natural compounds could help sub-Saharan Africa develop a traditional herbal medicine pipeline. Technology transfer in this area may also prove fruitful for firms in sub-Saharan Africa, for example Nigerian researchers started a local firm that uses the Neem plant (Box 9.2) and technology transferred from India to manufacture products locally for the Nigerian market.

Sub-Saharan Africa's collaboration with China and India has not yet reached its full potential, whereby firms are engaging in local innovation. However, our interviewees feel that learning from China and India's experience through collaboration can help them leapfrog into forming more innovative partnerships and potentially innovative biotechnology products targeted to the diseases of Africa. Firms in our study cite, they are interested in deepening their collaboration with China and India from marketing relations and engaging in joint innovation. They saw their existing collaborations as laying the groundwork for more innovative partnerships which would have a greater impact on the availability of locally relevant health products and on the development of health biotechnology in Africa. Still respondents emphasised the need to develop further policy mechanisms within Africa that specifically address engagement with China and India. Harnessing the impact South-South collaboration can have effectively rests with African governments and the wider African policymaking community. China and India have made efforts to formalise their trade policy and establish national frameworks for cooperation with Africa, but African countries have not constructed a coordinated approach or policy framework that would allow them to engage specifically with India or China. This is a critical step in leveraging benefits of collaboration for Africa, as such policies may help ensure equitable terms for African countries, help benefits fall into sectors where they are most needed, provide a framework for inputs from the broader community, and create a base on which subsequent relations can be built. Interviewees noted that it will be important for governments to take a bottom-up approach. That is to identify challenges they face in collaboration, and form policies that help them to overcome these basic hurdles. For firms, strengthening regional collaboration may help better position them to leverage African markets, and engage strategically with China and India. It would be of interest to examine policy options at the national, regional, or continental level and determine which approach may be best to carry forward.

9.6 Conclusions

Biotechnology is still a newcomer to most of sub-Saharan Africa. Typically their governments have only singled out the sector as being of importance to them in recent years. They have started institution building and some countries now have universities and research organisations doing impressive research in the health biotechnology sector and firms starting to engage in biotechnology. But resources are in short supply and in order to become more active contributors to research and innovation they need to strengthen capacity and infrastructure. Our research gives an indication that sub-Saharan Africa can harness South-South collaboration to gain this capacity and promote development and innovation of health biotechnology solutions appropriate for the African context. Historical ties and the conventional flow of research opportunities and funding from the North to the South have led to Northern domination in collaborative partnerships with Africa. This exclusive focus on collaborating with partners in the North is beginning to change and there is a growing recognition that partnerships with emerging economies have a lot to offer sub-Saharan Africa. The main messages from our case-study research on sub-Saharan Africa's health biotechnology collaborations with China and India show that:

Both China and India are becoming active collaborators with sub-Saharan Africa in the health biotechnology sector and are increasing their presence on the continent in this science-intensive field. Even though our mapping of China's and India's health biotechnology collaboration with sub-Saharan Africa shows only a moderate level of collaboration, in recent years there has been an upsurge of contracts promoting collaboration in science and technology in general. Further, health and biotechnology are typically priority areas among these countries' bilateral discussions and both China and India are allocating millions of dollars to support such collaborations. Time will tell if this is solely a part of their strategy to become global powers, win political favours, and access markets in Africa, or if it is an expression of South-South unity. Their ties with Africa are, thus, not confined to resource extraction, which is the frequent topic in Western media, but our research suggests they are increasing their presence within Africa's health biotechnology sectors. While criticism suggests that China and India's collaboration with much of Africa has been directed towards natural resources partnerships, our research indicates that steps have been taken for a much broader approach to collaboration, and both Chinese and Indian firms have started to form ties with Africa in the non-extractive sectors.

China and India play various roles in sub-Saharan Africa's health biotechnology but are particularly strong in capacity building. Both countries did play various roles in sub-Saharan Africa's health biotechnology development and both, for instance, had an active role in providing access to scientific infrastructure as well as the provision of relatively affordable health products appropriate to local health needs on the continent. Their roles in training and capacity building were particularly pronounced and are being strengthened by both China and India allocating millions of dollars for scholarships and fellowships in science and technology earmarked for African nationals. Their capacity-building efforts are also being reinforced by contributions from international initiatives and organisations, and both TWAS and ICGEB, were singled out by our interviewees as having played a considerable role in financially supporting their training and collaboration with China and India and facilitating collaboration by helping to identify potential collaborators. It appears from the research that they have played a large role in Kenya and Nigeria as compared to South Africa. The strong message from both researchers and entrepreneurs is that third party and philanthropic organisations can play a critical role in brokering South-South collaboration of Africa with China and India.

Shared health concerns foster sub-Saharan Africa's collaboration with China and India. Respondents involved in both firm and research collaboration cited the need for Southern countries to work together in addressing health concerns commonly neglected by Northern research agendas. Researchers and firms echoed that access to more appropriate technologies, common health concerns, and strong interest in traditional medicines were important areas motivating collaboration with China and India. These areas of commonality are particularly visible when looking at joint research publications which heavily focus on areas such as HIV, TB, malaria, and tropical diseases. A key message from our research is that collaboration with China and India has in many cases allowed researchers to orient collaborative work more closely to their local research agendas and has increased their ability to focus on research questions aligned to their local health needs. Respondents further supported this view citing China and India as important Southern players in developing appropriate, low-cost solutions that can contribute to improving of the lives of many Africans.

Collaboration with China and India in harnessing traditional medicine and local biodiversity is in high demand in sub-Saharan Africa. There is increasing potential for collaboration between China, India and sub-Saharan Africa in traditional medicine. Many researchers we interviewed emphasised that they see this as a primary avenue through which to collaborate with China and India especially on the development, advancement, and commercialisation of traditional knowledge. South Africans were particularly interested in understanding management of IP issues regarding traditional medicine, whereas Kenya and Nigeria were more interested in the development process of herbal medicines and technology transfer. Collaboration in this area has already begun and can be seen in such examples as in the case of transferring Neem plant technology to Nigeria and South African collaborations with India in the creation of a traditional medicines database. Such collaborations may become a way for Africa to advance its own capacity in the innovation, development, formulation, testing, and marketing of new herbal products while maintaining a focus on bringing affordable, culturally appropriate, health products to their populations. However, as in many cross-border collaborations, issues regarding standardisation and regulation remain key challenges.

Drivers for collaboration with China and India are not uniform in sub-Saharan Africa. Our case-study research showed that there are vast difference in the health biotechnology sectors in Kenya and Nigeria *versus* South Africa that result in differences in their collaboration with China and India. While Kenya and Nigeria collaborate from a resource poor position, South Africa collaborates with these two emerging economies more from an equal position. South Africa does not need to gain access to standard technologies and equipment in health biotechnology through collaboration with China as the country already has considerable strengths in this field. Kenya and Nigeria, however, need to collaborate in order to build up human capacity in this field and to gain access to scientific infrastructure and complete research projects. South Africa is also in the position to receive graduate students and fellows from China and India for specialised training in health biotechnology. It is still noteworthy that despite its strengths in health biotechnology, South African firms and researchers still consider China and India vital partners. Instead of wanting to gain access to technologies for harnessing traditional knowledge, South Africa wants to learn from China and India on how to develop, regulate, and commercialise products/services.

Our study of sub-Saharan Africa's health biotechnology collaboration with China and India shows that there are opportunities for Africa to strengthen capacity, knowledge production, and innovation through collaboration with other Southern countries. With continued government commitment to the development of the biotechnology sectors in Africa, and recognition of the contributions that building South-South linkages may have, India and China stand to make important contributions towards Southern-based knowledge production and innovation in Africa. South-South collaboration between these regions may provide an effective mechanism to boost the R&D of locally oriented products and develop knowledge for the benefit of public health in Africa.

References

Al-Bader, Sara, Sarah E. Frew, Insiya Essajee, Victor Y. Liu, Abdallah S. Daar and Peter A. Singer (2009). "Small but tenacious: South Africa's health biotech sector", *Nature Biotechnology* 27(5): 427-45.

Beijing Summit and the Third Ministerial Conference (2006). *Forum on China-Africa cooperation: Beijing action plan (2007-2009)*. Beijing: Government of the People's Republic of China.

Boshoff, N. (2010). "South–South research collaboration of countries in the Southern African development community (SADC)", *Scientometrics* 84(2): 481-503.

Broadman, Harry G. (2007). *Africa's silk road: China and India's new economic frontier*. Washington, DC: World Bank.

Cheru, Fantu and Cyril Obi (2010). *The rise of China and India in Africa: Challenges, opportunities and critical interventions*. London: Zed Books and Nordiska Afrikainstitutet.

Department of International Collaboration Ministry of Science and Technology (2010). *China-Africa science and technology partnership program (CASTEP)*. Accessed from: *http://www.cistc.gov.cn/englishversion/*

Dickson, D. (2003). "South-South collaboration picks up steam", *Science and Development Network*, November 17.

European Comission. *Community research and development information system* (CORDIS) Database of Statistics. Accessed from: *http://cordis.europa.eu.*

Fage, J. D. and W. Tordoff (2002). *A History of Africa*. New York: Routledge.

French, H. (2004). "China in Africa: All trade, with no political baggage", *New York Times*, August 8.

Frew, Sarah E., Rahim Rezaie, Stephen M. Sammut, Monali Ray, Abdallah S. Daar, and Peter A. Singer. (2007). "India's health biotech sector at a crossroads", *Nature Biotechnology* 25(4): 403-17.

Frew, Sarah E., Stephen M. Sammut, Alysha F. Shore, Joshua K. Ramjist, Sara Al-Bader, Rahim Rezaie, Abdallah S. Daar, and Peter A. Singer. (2008). "Chinese health biotech and the billion-patient market" *Nature Biotechnology* 26(1): 37-53.

Goldstein, A.E., P. Nicolas, H. Reisen and X. Chen (2006). *The rise of China and India: What's in it for Africa?.* Washington D.C.: Organization for Economic Cooperation and Development.

Government of Federal Republic of Nigeria (2005). *National policy for biotechnology*. Lagos.

Government of the People's Republic of China (2007). *Chinese scholarship council annual report 2007*. Beijing.

Government of the Republic of Kenya (2006). *Biotechnology policy for Kenya*. Nairobi: Ministry of Higher Education, Science and Technology.

Government of the Republic of South Africa (2007). *Technology innovation agency bill: section 75*. Pretoria.

———. (2006). *A national biotechnology development policy*. Pretoria.

Hassan, M.H.A. (2007). "Building capacity in the life sciences in the developing world", *Cell* 131(3): 433-6.

India, Brazil, South Africa (IBSA) Summit (2007). *New Delhi Ministerial Communiqué.* New Delhi: India.

———. (IBSA) Summit (2005). *Cape Town ministerial communiqué*. Cape Town: South Africa.

———. (IBSA)-Trilateral agreement (2003). Accessed from: *http://www.ibsa-trilateral.org*

India-Africa Forum Summit (2008a). *Delhi Declaration.* New Delhi: Government of India.

———. (2008b). *Africa framework for cooperation.* New Delhi: Government of India.

Jeenah, M. and A. Pouris (2008). "South African research in the context of Africa and globally", *South African Journal of Science* 104(10).

Jia, Hawk (2006). "African science to benefit from china trade deal", *Science and Development Network*, November 6.

Kagame, H.E. Paul (President of the Republic of Rwanda) (2006). "The importance of science and technology in Africa", Paper presented at The Royal Society, London.

Kaplinsky, Raphael and Mike Morris (2009). "Chinese FDI in sub-Saharan Africa: Engaging with large dragon", *European Journal of Development Research Special* 24(1).

Louët, S. (2006) "Rainbow biotech—South Africa's emerging sector", *Nature Biotechnology* 24(11): 1313-20.

Manji, Firoze and Stephen Marks (2007). *African perspectives on China in Africa.* Nairobi and Oxford: Fahamu.

Mboya-Okeyo, T., R.G. Ridley, S. Nwaka and ANDI Task Force (2009). "The African network for drugs and diagnostics innovation", *Lancet* 373(9674): 1507-8.

McBride, Edward (2008). "The new colonialists: A ravenous dragon: A special report on China's Quest for Resources", *Economist*. March 13–21.

Ministry of Foreign Affairs of the People's Republic of China (2010). *Forum on China Africa cooperation*. Accessed from: *http://www.fmprc.gov.cn*

Mokoena, R. (2007). "South-south co-operation: The case for IBSA", *South African Journal of International Affairs* 14(2): 125-45.

Motari, M., U. Quach, H. Thorsteinsdóttir, D.K. Martin, A.S. Daar, and P.A. Singer (2004). "South Africa–blazing a trail for African biotechnology", *Nature Biotechnology* 22: 37–41.

Moyo, D. (2008). *Dead aid: Why aid is not working and how there is a better way for Africa*. New York: Farrar, Straus and Giroux.

Naidu, S. (2007). "The forum on China-Africa cooperation (FOCAC): What does the future hold?", *China Report* 43(3): 283.

New Partnership for Africa's Development (NEPAD) (2006). *Africa's science and technology consolidated plan of action*. Pretoria.

Nwaka, S., T.B. Ilunga, J.S. Da Silva, E.R. Verde, D. Hackley, R. De Vré, T. Mboya-Okeyo, and R.G. Ridley (2010). "Developing ANDI: A novel approach to health product R&D in Africa", *PLoS Medicine* 7(6).

Organisation for Economic Co-Operation and Development (OECD) (2006). *South-south trade: Vital for development.* Paris: Organisation for Economic Co-Operation and Development (OECD).

Pouris, A., and A. Pouris (2009). "The state of science and technology in Africa (2000–2004): A scientometric assessment", *Scientometrics* 79(2):297-309.

Puri, Lakshmi (2007). "IBSA: An emerging trinity in the new geography of international trade", Paper presented at United Nations Conference on Trade and Development, New York.

Rotberg, Robert (2008). *China into Africa: Trade, aid, and influence*. Washington, D.C.: Brookings Institution Press and World Peace Foundation.

Sahn, E., P. Dorosh, and S. Younger (1999). *Structural adjustment reconsidered: Economic policy and poverty in Africa*. London: Cambridge University Press.

Schwab, K., Robert Zoellick and Donald Kaberuka (2009). *The Africa competitiveness report 2009*. Washington World Bank.

South Center (2006). "IBSA: Assessing the trilateral south-south initiative", *South Bulletin* (122): 183-185.

The Sharm El Sheikh Summit and 4th Ministerial Conference (2009). *Forum on China-Africa cooperation Sharm el Sheikh action plan (2010-2012)*. Sharam El Sheikh, Egypt.

The World Bank (2010). *World bank data catalouge: Population ranking tables*. Accessed from: *http://data.worldbank.org/data-catalog*

Toyoshima, Toshihiro, Yutaka Yoshino and Chad Leechor (2004). "Patterns of Africa-Asia trade and investment: Potential for ownership and partnership", Washington: World Bank.

United Nations Conference on Trade and Development (2009). *UNCTAD handbook of statistics*. Accessed from: *http://stats.unctad.org*

———. (2004). *The biotechnology promise: Capacity in the bioeconomy*. New York: United Nations Conference on Trade and Development.

United Nations Development Programme (2009). *South report 2009: Perspectives on South-South cooperation for development*. New York, NY: Special Unit for South-South Cooperation.

Wambebe, Charles (2007). *NIPRISAN case Nigeria - A report for genbenefit*. Lancashire: University of Central Lancashire.

World Trade Organization (2008) *International trade statistics*. Geneva: WTO Publications.

Zeleza, T. (1997). *A modern economic history of Africa: The nineteenth century*. Nairobi: East African Educational Publishers.

10 Promoting South-South Collaboration Revisited

AUTHORS: *Halla Thorsteinsdóttir, Andrew Kapoor, Tirso W. Saenz, Maria Carlota de Souza Paula, Heba Maram, Magdy A. Madkour, Lexuan Li, Wen Ke, Victor Konde, Zhang Jiuchun, Marwa G. Elwakil, Nefertiti El-Nikhely, Sachin Chaturvedi, Sahar Aly*

The voyage taken in this book has explored South-South collaboration from different angles. We have mapped the levels and key characteristics of health biotechnology collaboration involving both researchers and entrepreneurs and looked in detail at diverse bi-national collaboration initiatives involving 13 developing countries from different parts of the world. During this journey we have interviewed around 350 experts in developing countries who have had direct experience of collaborating with scientists and entrepreneurs in other developing countries and/or had experiences in promoting or otherwise influencing South-South collaboration. In this chapter, we want to take stock of our key learning about South-South collaboration. We want to present recommendations based on the collective advice of our interviewees and our further analysis of the data from this study. In our recommendations we highlight what can be done to strengthen collaboration amongst developing countries in health biotechnology, and enhance its contribution towards scientific capacity, economic growth, and health. Our research is focussed on collaboration in the health biotechnology sector, but many of the experts we interviewed, and policies and programmes we reviewed, incorporated wider applicability to science and technology in general.

10.1 Is South-South collaboration rhetoric or reality?

Our mapping of health biotechnology collaborations showed that South-South collaboration has become a reality in the health biotechnology sector, but it is not evenly experienced by researchers *versus* firms or by the different countries. South-South collaboration is relatively common amongst health biotechnology firms in developing countries, and the results from our survey discussed in Chapter 3 show that one in every four firms are collaborating with other developing countries. We also saw that a large proportion of the collaboration was between the leading developing countries in the health biotechnology field, and therefore was typically trans-continental in nature. China collaborates mainly with Brazil and India, and India has close linkages with South Africa. In addition there was some regional collaboration flourishing, for instance, between South Africa and other sub-Saharan countries and Cuba and Brazil. When we inquired about what activities were involved in the South-South collaboration, we found out that most of the collaboration was in marketing and dissemination of their products within each other's markets. This finding is consistent with the observation discussed in Chapter 1 that South-South trade in general has become a significant portion of developing countries' trade and expanded immensely in recent years (UNCTAD, 2009). Our case-study research on the bilateral collaborations, particularly the research discussed in Chapter 9, also supports the notion that the new multinational firms headquartered in developing countries constitute a part of this collaboration (Battat and Aykut, 2005). This was the case for Indian firms, such as Dr. Reddy's Laboratories (Hyderabad, India) and Cipla (Mumbai, India). As such the intensity of South-South firm collaboration in health biotechnology partly reflects an increased extent of South-South foreign direct investment.

While South-South entrepreneurial collaboration seems to have become relatively common, collaboration between researchers in developing countries is generally rarer. When we looked at all low- and middle-income countries, less than 10 per cent of their co-authored papers were with each other and their co-authored papers did not seem to be increasing (Chapter 2). However, a larger emphasis seems to be placed on South-South collaboration by most of the focal countries in this study.

In 2006–2009, for instance, 25 per cent of South Africa's internationally co-authored papers in health biotechnology were with other developing countries, 20 per cent of Egypt's, and 19 per cent of Brazil's. Moreover, their emphasis on South-South collaboration appeared to be increasing. Only China's emphasis on South-South collaboration remained small. Considering that China is such a large contributor to health biotechnology publications, and ranks second globally in terms of number of health biotechnology publications, its relatively limited emphasis on South-South collaboration colours the overall levels of such collaboration. Even though China's collaboration with developing countries on the whole remains limited, our study on China's collaboration with sub-Saharan Africa discussed in Chapter 9 shows that its research collaboration with several African countries appears to be building up steam. Our research on South-South collaboration discussed in Chapter 2 also showed that Brazil has the highest number of South-South co-authored papers in the health biotechnology field and has particularly frequent linkages with Argentina, but also with other emerging economies such as China and India. India and China also have relatively frequent collaborations with each other.

From our analysis, we therefore conclude that South-South collaboration is far from being rhetoric and has become a reality in the health biotechnology sector. However, a closer look at the pattern of collaboration shows that it is uneven. Firms are embracing it and setting up collaboration even across continents, but collaboration on the research side does not seem to flourish to the same extent as on the entrepreneurial side. There are important exceptions to the latter observation as most of the focal countries in our study had a considerable emphasis on South-South research collaboration which was increasing over time.

10.2 What are the key opportunities arising from South-South collaboration?

Our study on South-South collaboration identified various opportunities that the collaboration can offer. The key ones discussed here are opportunities to extend capacity in health biotechnology between developing countries and strengthen the ability to provide affordable health products appropriate to developing countries' health needs.

Through South-South collaboration capacity in health biotechnology can be extended between developing countries

As we saw in Chapter 2, capacity in health biotechnology is highly unequal between developing countries. While some countries such as Brazil, China, Cuba and India have research and innovation capacity on par with many developed countries, others lag far behind and do not have the capability to carry out rudimentary research in this field. All the countries that are relatively weak in the health biotechnology field that we studied in this research project stressed the importance of South-South collaboration contributing towards capacity building in their countries. This was a relatively common theme in our African interviews where there was great scope to build up health biotechnology capacity in countries such as Egypt, Kenya, Nigeria and Zambia through South-South collaboration (see Chapters 6, 8 and 9).

In some of the cases it was stressed that the South-South collaboration allowed for a more appropriate level of technology than collaboration with Northern countries that often included the use of more expensive technologies. As health biotechnology is a highly specialised field, there is scope for capacity building in specialised techniques even amongst countries that have highly functional health biotechnology sectors. For instance collaboration between the Kunming Institute of Zoology (Kunming) of the Chinese Academy of Sciences, the Sanjay Gandhi Postgraduate Institute of Medical Sciences (Lucknow), and the North Bengal University (Siliguri), discussed in Chapter 5, involved gaining access to Chinese sequencing infrastructure and expertise and helped India to build further capacity in genomics.

Even though some countries are predominantly in the role of providers of capacity, they rarely engage in South-South collaboration without it being of some value to them as well. The world is small and the problems of one country affect other countries near and far. A single collaboration initiative is typically driven by multiple reasons. For example, when South Africa is providing capacity in health research to neighbouring SADC countries they are likely to be contributing to strengthened capacity to deal with epidemics and enhance SADC countries' potentials to contain and address health threats within their borders. They thereby both fulfil their global citizen's role and strengthen their ability to address infectious diseases. When China contributes to strengthened sequence capacity in India they may be obtaining access to samples from India that can be an important contribution to their own research. They are further strengthening the potential for them to have mutually fruitful collaboration with India in the future.

The capacity built through South-South collaboration also included competence in harnessing biodiversity or traditional knowledge. We saw this both in Asia and widely in Africa. As discussed in Chapter 9, NIPRID (Abuja, Nigeria) has developed a drug against sickle cell anaemia based on traditional knowledge and has, for instance, collaborations with a number of institutions in India and China including the Council of Scientific and Industrial Research (New Delhi, India), the University of Lucknow (Lucknow, India), and Zhejiang University (Hangzhou, China). We also saw collaboration between the Chinese firm SH-IDEA Pharmaceutical Company (Yuxi, China), the Kunming Institute of Botany (Kunming, China), and Thailand's Ministry of Public Health (Bangkok, Thailand) (Chapter 5) on clinical trials of a HIV/AIDs treatment based on Chinese traditional medicine and local biodiversity. In general we saw a demand for the experience that China and India have in specialised technical knowledge for harnessing biotechnologies, but also a demand from their collaborators for knowledge on how to manage the development process, how to protect the intellectual property rights (IPRs) embedded in the traditional knowledge or local biodiversity, and how to regulate the development of health products based on this type of knowledge/ resources. South African policymakers, for example, stressed the benefit of learning from the Indian experience of managing their innovation based on traditional knowledge and biodiversity.

South-South collaboration can strengthen the ability to focus on local health problems and produce more affordable health products

There was a strong message in our interviews with researchers that South-South collaboration enabled research that was closely aligned with local health problems and we heard it to some degree from all our bilateral case studies. By working together, researchers were able to gain access to each other's expertise, samples, research infrastructure, and other resources and enhance their ability to react to their own health problems. Frequently our interviewees offered comparison to collaboration with Northern researchers and felt this type of collaboration was not as clearly focussed on needs in developing countries. The strengthened ability to focus on local health problems fits well with the general rationale for South-South collaboration discussed in Chapter 1, where ability to address shared problems has been a strong driver generally for collaboration amongst developing countries. It was also evident that the enhanced opportunity to focus on local health problems was experienced by both researchers from the emerging economies and the countries weaker in health biotechnology. For instance, researchers in China and India highlighted this opportunity as well as researchers in Egypt, Kenya and Nigeria.

Both researchers and entrepreneurs we interviewed also stated that South-South collaboration made it possible to gain access to and develop affordable health solutions. In the latter cases we observed that South-South collaborations involving marketing and distribution led to trading of more affordable health products than comparable alternatives on the world market. For example, our interviewees argued that treatments for hepatitis, HIV/AIDs, malaria, and TB are now available in several African countries through South-South collaboration with generic producers from China and India. Firms such as Holley-Cotec (Beijing, China), Ranbaxy (Gurgaon, India), and Shanta Biotechnics (Hyderabad, India, now acquired by Sanofi-Aventis, France) are providing lower cost alternatives, thereby providing more affordable health products to African countries. Technology transfer arrangements between developing countries have also increased the availability of affordable health products. For instance, as discussed in Chapter 4, the Brazilian health system could purchase more affordable EPO, from a Brazilian supplier Bio-Manguinhos (Rio de Janeiro) because of technology transfer from CIMAB (Havana) in Cuba. Another opportunity of this arrangement was the ability of developing countries to become more self-sufficient in the supply of necessary health products and to lessen their reliance on imports. This opportunity was also stressed as being afforded by Egypt's collaboration with China in developing recombinant insulin discussed in Chapter 6, where the technology was transferred from the Chinese company Dongbao (Shanghai) to VACSERA in Giza, Egypt. The importance of self-sufficiency can be politically valuable in an unstable world.

The opportunity for the South-South collaboration to lead to affordable health products that are closely aligned to developing countries' health needs can also be beneficial to countries that do not directly take part in the collaboration. We identified a case where Brazil and Cuba have been collaborating to produce meningitis A vaccine for an outbreak of *Neisseria meningitidis serogroup A* in Africa (Chapter 4). Neither Brazil nor Cuba have had outbreaks with this serogroup of meningitis. Bio-Manguinhos in Brazil and the Finlay Institute (Havana) in Cuba used their respective strengths in the development and manufacturing process to provide African countries with a vaccine to prevent further spread of the disease. According to the Finlay Institute, between 2007 and 2009, 19 million doses were produced and distributed in Burkina Faso, Ethiopia, Mali and Nigeria. The vaccine's price is much lower than on the international market, and lower than would be possible without cooperation. Neither firm alone would have been able to respond as quickly and efficiently to this health threat as they managed to do through their collaboration.

From the above points we see that although developing countries are becoming more diverse over time and the conditions in the emerging economies are becoming more similar to the conditions in developed nations, there is still a cluster of common interests in the health biotechnology sector that unite developing countries. Developing countries have considerable potential to learn from each other, extend capacity in science-intensive fields such as health biotechnology, and to strengthen research and innovation on their shared health problems.

10.3 What makes South-South collaboration possible?

Our research identified several factors and conditions that encourage South-South collaboration. Governments in developing countries have a clear role in promoting collaboration between developing countries. The Brazilian government has, for example, been successful in cultivating collaboration with developing countries and has placed an emphasis on South-South collaboration in different domains (Chapter 4). It has established trade relations between developing countries but has also driven South-South collaboration in scientific fields. It has played a strong regional role in Latin America but also has promoted more far reaching South-South collaborations both with Portuguese speaking African countries and the emerging economies. The role of governments in promoting collaboration between Brazil and Argentina is clear. The governments in the two countries established the CBAB/CABBIO fund in 1986 that continues to support joint research and training activities between the two countries. Brazil and Argentina now have the largest number of co-publications in health biotechnology between any two developing countries. This, therefore, supports the notion that governments can stimulate South-South collaboration by investing resources for collaboration.

The Southern governments have, however, only played a small role in directly promoting collaboration between health biotechnology firms with only seven per cent of the firms we surveyed reporting a role by governments in initiating the collaboration. The firms are mainly collaborating because it makes business sense and facilitates their marketing efforts. Some of our interviewees reported a challenge in identifying appropriate partners in other developing countries and initiating the collaboration. For small biotechnology firms it can be a difficult task to find enough details about potential partners and to build trust between the firms, so there is certainly scope for governments to help them take these initial steps in collaboration. There is also a scope for governments to encourage South-South collaboration to a larger extent than they currently do to lead to the development of new health products and services. Our survey of health biotechnology firms showed a limited focus on research and developmental efforts by firms engaged in South-South collaboration. There is, however, inadequate funding available for South-South developmental efforts and governments generally do not target these kinds of collaborations. For South-South collaboration to have a stronger innovation record, there is thus a need for more funding for their joint developmental efforts. These funds do not have to come from governments alone and non-governmental and private capital, such as venture capitals funds, could step in and support South-South firm collaboration. They may, however, need some incentives and directions from government to do so.

Our research also showed that international associations and international organisations played an important role in making South-South collaboration in health biotechnology possible. TWAS had a considerable role in promoting South-South collaboration. It has dedicated funds to facilitate both scientific and educational exchanges that several of our interviewees stressed played an important role for capacity building, particularly in African counties. The ICGEB is a part of the UN system and has branches both in New Delhi, India and Cape Town, South Africa. It has 39 affiliated institutions (one in each member country) that form a South-South network. It plays a significant role in research and training activities for improving biotechnology R&D both in health and agriculture. Regionally some formal networks have played a catalyst role. For instance, AMANET has played a role in building capacity and promoting African-led R&D specifically in the area of malaria. It also has a strong role in promoting infrastructure for clinical testing of new malaria vaccine candidates. TDR, co-sponsored by UNICEF, UNDP, World Bank and WHO, is also a South-South network that plays a role in collaboration on infectious diseases. These are just some of the examples of networks in place that support South-South collaboration.

A good example of the pivotal role of international organisations is the role the WHO played in the collaboration between Brazil and Cuba in producing a vaccine for the meningitis belt in Africa discussed above. WHO reacted to the meningitis outbreak in Africa by accessing the status and production capacity of polysaccharide vaccine manufacturers worldwide (WHO, 2007a and 2007b). This examination identified Bio-Manguinhos, in collaboration with the Finlay Institute, as the most suitable suppliers and WHO helped initiate their collaboration. The joint effort permitted a fast positive response to WHO's call, and distribution of this vaccine by WHO in various African countries. To support the joint efforts of the Finlay Institute and Bio-Manguinhos, the regulatory agencies in Brazil and Cuba started working together. Because of the WHO's role in the Africa meningitis project—including prequalification of the meningitis AC vaccine—the Bio-Manguinhos and the Finlay Institute were given an extra push to collaborate and had more funding available for this purpose than they would have had otherwise.

As we can see, South-South collaboration is made possible by contributions from various organisations from different parts of the developing world as well as international organisations. There is no single simple recipe that will work to promote South-South collaboration but coordinated efforts of various organisations are key to successful collaboration.

10.4 What hinders South-South collaboration?

South-South collaboration is seriously hampered by a lack of dedicated resources. To date, governments in developing countries have not allocated significant resources to fund South-South collaboration. In general, the message was that resources for South-North collaborations were more easily available and typically provided by high-income countries. The availability of resources, therefore, provided a stronger impetus for South-North than South-South collaborations. While South-North collaboration in the health field is certainly of great importance to developing countries, many of our interviewees expressed concern that there was a risk that it diverted emphasis away from developing countries' needs as discussed above.

However, despite the strong economic growth of many developing countries and biotechnology's prominence on the South-South collaboration agenda, we could identify only one fund that is dedicated to fund biotechnology collaborations, the CBAB/CABBIO fund between Brazil and Argentina (initiated in 1986) which has significantly promoted collaboration between the two countries. As discussed above Brazil and Argentina thus now have the largest number of co-publications between any two developing countries. Governments can therefore stimulate South-South collaboration by investing resources in the collaboration. Because the IBSA initiative has prioritised research on several health problems it does, in practice, encompass a considerable focus on South-South health biotechnology collaborations. The IBSA Trust Fund was established in 2004 for promoting joint research between Brazil, India and South-Africa focussed on HIV/AIDs, malaria, and other diseases. As it is a more recent development, its impact on South-South research collaboration has been limited so far.

Another key challenge that our research identified is limited effects of the collaboration on innovation. The collaboration may involve capacity building, but the new capacity is not necessarily used in applied settings and may therefore not contribute to research, health improvements, or economic development. For instance, the collaboration often involves providing research training for researching health problems in developing countries. Even though the training may be highly relevant to the health problems of the country lagging behind scientifically, the trainee may not have the facilities to continue his/her work after receiving the training, as there is limited research infrastructure in the receiving country. It thus reflects how the collaboration is not aligned with the prioritisation or the innovation systems in the nations lagging behind and thus has only limited impact. The fact that there are systemic deficiencies in scientifically lagging countries, therefore, limits the interaction potential within the innovation systems in the participating countries and thwarts South-South collaboration.

A further challenge for South-South collaboration is the limited linkages of those engaged in South-South research collaboration with firms or other entrepreneurial organisations in their countries that can develop and commercialise health products and services. For instance, the Latin American collaboration on Chagas diagnostics, discussed in Chapter 4, resulted in knowledge that could be applied to develop a diagnostics kit for Chagas disease. The commercialisation of the Chagas diagnostic kit is however, not being carried out by a Latin American firm, collaborating with the research group. Instead the US firm Chembio Diagnostic Systems is commercialising a diagnostic kit based on the Latin American collaboration. Tighter linkages of South-South research with entrepreneurial agents could strengthen Southern commercialisation of biotechnology products.

For collaboration involving manufacturing or development of health products, it could be a challenge for the collaboration to deal with regulatory systems in both participating countries. Some of the countries had immature regulatory systems which add a special difficulty to bringing health products on the market. In other cases, vastly different regulatory processes hampered the collaboration. Some tests had to be repeated and/or different types of information were required. Other challenges commonly cited were the excessive time and the high cost of moving products or ingredients for the production across international borders. Such transactions were particularly difficult after the heightened security demands following the September 11, 2001 attack in the United States.

The challenges discussed so far in this section represent misalignments between the innovation systems in the participating countries both in the realms of research and entrepreneurial collaborations. In order to strengthen the impact of the collaboration, it is important to look at South-South collaboration as interactions between two systems in different developing countries. If country A provides training to country B, it is of key importance that the training is relevant to the priorities of country B, and that the trainee has the systemic support to continue to work in the area in country B. Instead of looking at the collaboration as isolated training, it is crucial to consider how the training fits into the health and innovation systems in both countries. By doing this, national ownership of the expertise is enhanced and the collaboration is better aligned with the priorities of both countries.

In entrepreneurial collaboration system alignments are also needed. We can, for example, see from the discussion above that there is a need for system alignments of regulatory processes. When firms in two countries are engaged in joint development of health products, it can delay development to have to work with regulatory systems in two countries. If the regulatory systems have the opportunity to collaborate, exchange information about each other's requirements, and align their processes, the development is likely to be less challenging. In the case of the Brazil-Cuba collaboration for meningitis A vaccine in Africa, the fact that the WHO initiated the collaboration and supported it, allowed the regulatory systems in the two countries to collaborate. As a result, the collaboration was smoother and it was quickly able to produce a cost-effective vaccine that met health needs in Africa. Systemic alignment of entrepreneurial collaboration can thus be of key importance for effectiveness.

10.5 What can be done to strengthen South-South collaboration?

We asked our interviewees to give us advice on what can be done to strengthen South-South collaboration. We observed a high degree of consensus in their advice across the different countries we examined. Below we list the key recommendations presented by our interviewees augmented with our further analysis on the topic.

10.5.1 Increase financial resources that support the collaboration

Not surprisingly there was a consensus in all the countries that we studied that in order to strengthen South-South collaboration more financial resources were needed. The collaboration between researchers was frequently halted because of lack of resources. There was a sense that governments needed to back up verbal commitments with financial resources funding South-South collaboration. Despite economic growth in many of the emerging economies they have made inadequate efforts to establish funds that could support health biotechnology research and developmental efforts. Scientific and technological cooperation agreements were typically stressed as being important as a first step and developing countries needed to sign more agreements with each other, but financial resources needed to follow. As was argued in Chapter 7, without more financial resources there is a risk that poorer developing countries will be left out of South-South collaboration and the collaboration will be limited to the *nouveau riche* who can fund their collaboration themselves. This development would go against the spirit of South-South collaboration expressed from its onset and increase the gap between the scientifically advanced and lagging developing countries.

It was also a common theme that many more funds were streaming from the North, which encouraged South-North collaboration. Some co-funding schemes have also started to appear between funding organisations in the North and Southern governments. For example, the German organisation DAAD now has an agreement with the Ministry of Higher Education and Scientific Research in Egypt to co-fund doctoral scholarships for Egyptian students to study in Germany. While this arrangement can be important for gaining access to necessary training opportunities, it does attract resources from developing countries' governments for collaboration with the North. Northern funding organisations are more likely to initiate this type of arrangement than Southern funding organisations as the former have more resources. Governments in developing countries need to carefully evaluate if co-funding schemes with Northern governments are the best way for them to use their limited resources. An alternative would be to devise a comparable co-funding scheme with Southern funding sources where, potentially, training is less expensive and can therefore be extended to more candidates.

Alternatively and to extend resources for South-South collaboration, some interviewees also stressed the importance of South-South-North collaboration where developing countries' governments and donor organisations from the North should work together to promote South-South collaboration. The knowledge sharing and capacity building could mainly be between developing countries, but resources and some technologies could come from the North. This may assist in harnessing the appropriate learning between developing countries and the technological and financial strengths of the North. But care has to be taken that this is not just fund-driven collaboration pushed from the North as was discussed in Chapter 8.

10.5.2 Prioritise areas of collaboration

There was also a theme that governments should prioritise particular areas to support. Without prioritisation the limited support would be spread too thin and not have much impact. This had to be a joint process involving all participating countries. It should take into account the needs of the participating countries but also the available capacity. Considering that developing countries typically share health problems and increasingly differ in their capacity in health biotechnology, there is considerable scope to identify promising priority areas. This is a practice that Brazil, for example, has undertaken in both its collaboration with Argentina and Cuba (Chapter 4) where shared health problems have been identified and guide the collaborations. Still, both Argentina and Cuba have had problems in allocating enough resources to meet their commitments. Prioritisation, therefore, does not automatically result in the intended impacts but needs to be followed up with dedicated funding in areas of high priority in developing countries.

Related to prioritisation, there were concerns in some of the countries that the budding South-South collaboration initiatives did not build upon previous experiences. Chapter 7, for instance, discussed the concern of some of the Indian interviewees that even though they had been engaged in fruitful South-South collaboration for years with limited financial support, they did not manage to get support from the newly established funds in India earmarked for South-South collaboration. Instead of supporting researchers that had, sometimes for years, taken part in South-South collaboration, support was just given to the most established researchers. The latter group typically had had collaboration with the North and were only initiating South-South collaboration to gain access to financial resources. They were engaged in fund-driven collaborations that are only likely to last while funds are available. Instead of only supporting the most established scientists, it is preferable to support more genuine South-South collaboration.

10.5.3 Include support for entrepreneurial collaboration

A further theme in resource allocation was the need not only to support training and research activities, but also to support collaboration activities of firms or entrepreneurial organisations involving joint development of health products and services. In the health biotechnology field, this would include joint engineering activities as engineering input is necessary for scaling up of biotechnological processes and other processes necessary for the development of health products. It would strengthen the limited innovation activities involving South-South collaboration and if successful make it more sustainable. Our survey showed that most of the firms that have started to focus on joint R&D activities are also involved in marketing collaboration. They are likely to start by pursuing only marketing activities and as trust has built up they start to deepen their relationships and attempt joint innovation activities. Even though innovation focus has strengthened in the health sector in developing countries, a large proportion of the firms are solely focussed on generics production and no real developmental activities.

A related recommendation is to promote technology transfer activities between developing countries where a country that possesses technology in demand helps a collaborating country master the technology and set up conditions to apply the technology. This can involve licensing the technology to a firm in the collaborating country. Technology transfer between developing countries can be a promising strategy to gain access to technologies that are more affordable and appropriate to developing countries' needs than the technologies from developed countries. It can thus lead to more targeted efforts in extending health biotechnology capacity to countries that are weak in this field, and start bridging the divide that exists between developing countries in health biotechnology. It was also stressed that it was important to encourage learning between developing countries on how to manage technology development, such as the development of intellectual property portfolios. There is further a large scope for developing countries to learn from each other how to harness their traditional knowledge and biodiversity. The likely pay-off of encouraging South-South entrepreneurial collaboration could be affordability and attention to local health needs. With an increased innovation focus, developing countries can leverage their individual strengths in this field and increase the pool of resources to address their shared problems.

10.5.4 Facilitate communications and the sharing of resources

Facilitating communication and sharing of information and other resources were also the recommendations echoed by many of our interviewees. In some of the countries we studied there were huge cultural and linguistic differences that impeded collaboration. As we discussed earlier in this chapter it could also be challenging for developing countries' firms in particular to find information about potential partners. There is therefore a need to bring researchers and firms together across developing countries and organise events and information sharing sessions in prioritised areas. Suggestions were also made to make better use of internet channels to exchange information, and for instance, have databases of available equipment in developing countries that could be shared with other countries. The desire for active sharing of resources went beyond equipment and many interviewees suggested that governments should also ease transport of biological materials and other necessary ingredients for health biotechnology, between countries. As discussed above the problem of exporting material between developing countries seriously hinders collaboration between them, particularly after the September 11 attacks in the United States. Pre-qualification of certain partners or partners' collaboration activities should be considered in order to facilitate and shorten this process.

10.5.5 Integrate South-South collaboration

For South-South collaboration to be effective it cannot be a stand-alone activity but rather needs to be a part of developing countries science, technology, innovation, and health promotion plans. Collaboration should not be seen as occurring between just two individual scientists or firms, but rather should take into consideration the wider institutional bodies in the participating countries. Promoting innovation based on the collaboration involves an understanding of the institutional environment in which the partnerships are embedded; it entails the identification of bottlenecks and the re-calibration of the system to overcome them. Research and training activities, for example, need to fit well with the needs in the receptor countries and the available infrastructure. When the collaboration is integrated in the participating countries' innovation systems, it has enhanced chances to have impacts. There should, therefore, be a greater dialogue between developing countries, and greater efforts should be made to promote the alignment of their health, science, and innovation systems in order for South-South collaborations to have the intended impacts. In order to strengthen the impact of the collaboration, it is important to look at South-South collaboration as interactions between two systems in different developing countries. The calibration for innovation then involves alignments between the systems in two or more participating countries so that they work most effectively together. This calibration has to include diverse parts of the innovation systems, and for development of health products and services, cooperation between the participating countries' regulation systems can be of vital importance. In order to smoothen the process of innovation they may need to consider how their regulatory offices can work together to make the process of cross-border innovation easier and faster.

10.6 Building bridges and promoting global health

In this book, we have presented information on the extent to which developing countries are involved in South-South collaboration and the key opportunities arising from this collaboration. We have also highlighted what is holding South-South collaboration back and limiting the realisation of its opportunities and further presented some key recommendations on what can be done to cultivate collaboration amongst developing countries and strengthen its impacts. Most of the recommendations are directed at organisations in developing countries. In a globalised world concerted efforts are, however, called for in order to harness the benefits of South-South collaboration. By relying on diverse countries' respective strengths, we can reinforce the research and developmental efforts on addressing health problems of developing countries and lead to availability of more affordable health products/services. As a result, South-South collaboration may be able to supply health products that reach more deprived people in developing nations. Since global health inequities are immense, there is full reason to harness this opportunity by both developing countries and the international community. International organisations, philanthropic organisations, donors, and governments in the North and the South should utilise South-South collaboration as a means for promoting development and global health. Through concerted efforts and well-aligned innovation systems, global efforts can sow the seeds of more South-South collaboration and harness it for those in need.

References

Battat, J. and D. Aykut (2005). *Southern multinationals a growing phenomenon*. The World Bank and International Finance Corporation.

UNCTAD (2009). *South-South trade: The realty check.* Geneva: United Nations Conference on Trade and Development. *Available at http://r0.unctad.org/ditc/tab/publications/DITC-TAB-2008-1-final-17MARCH2009.pdf*

WHO (2007a). *Meningitis in Africa: Hundreds of thousands vaccinated.* World Health Organization.

———. (2007b). *WHO coordinates response to meningitis outbreaks in four African countries.* Geneva: World Health Organization. Available at *http://www.who.int/mediacentre/news/notes/2007/np12/en/index.html*

About the Editor/Contributors

Editor and Contributor:

Halla Thorsteinsdóttir, Dalla Lana School of Public Health, University of Toronto, Toronto, Canada.

Contributors:

Sidar Abdusamad, Sekunjalo Health Care, Cape Town, South Africa.

Sahar Aly, Center for Special Studies and Programs (CSSP), Bibliotheca Alexandrina, Alexandria, Egypt.

Éric Archambault, Science-Metrix, Montréal, Canada.

David Campbell, Science-Metrix, Montréal, Canada.

Jennifer Chadder, McLaughlin Rotman Centre for Global Health, University of Toronto & University Health Network, Toronto, Canada.

Sharon Chakkalackal, McLaughlin Rotman Centre for Global Health, University of Toronto & University Health Network, Toronto, Canada.

Sachin Chaturvedi, Research and Information System for Developing Countries (RIS), New Delhi, India.

Grégoire Côté, Science-Metrix, Montréal, Canada.

Jan E. Cooper, McLaughlin Rotman Centre for Global Health, University of Toronto & University Health Network, Toronto, Canada.

Abdallah S. Daar, Dalla Lana School of Public Health, University of Toronto, Toronto, Canada; McLaughlin Rotman Centre for Global Health, University Health Network & University of Toronto, Toronto, Canada.

Marwa G. Elwakil, Center for Special Studies and Programs (CSSP), Bibliotheca Alexandrina, Alexandria, Egypt.

Nefertiti El-Nikhely, Center for Special Studies and Programs (CSSP), Bibliotheca Alexandrina, Alexandria, Egypt.

Chen Guang, Institute of Policy and Management, Chinese Academy of Sciences, Beijing, China.

Zhang Jiuchun, Institute of Policy and Management, Chinese Academy of Sciences, Beijing, China.

Andrew Kapoor, McLaughlin Rotman Centre for Global Health, University of Toronto & University Health Network, Toronto, Canada.

Wen Ke, Institute of Policy and Management, Chinese Academy of Sciences, Beijing, China.

Victor Konde, University of Zambia, Lusaka, Zambia.

Lexuan Li, Department of Human Resource Management, China Women's University, Beijing, China

Michelle Li, McLaughlin Rotman Centre for Global Health, University of Toronto & University Health Network, Toronto, Canada.

Magdy A. Madkour, Arid Lands Agricultural Research Institute, Ain Shams University, Cairo, Egypt.

Heba Maram, Center for Special Studies and Programs (CSSP), Bibliotheca Alexandrina, Alexandria, Egypt.

Christina C. Melon, McLaughlin Rotman Centre for Global Health, University Health Network & University of Toronto, Toronto, Canada.

Maria Carlota de Souza Paula, Centre for Sustainable Development, University of Brasilia, Brasilia, Brazil.

Monali Ray, McLaughlin Rotman Centre for Global Health, University of Toronto & University Health Network, Toronto, Canada.

Tirso W. Sáenz, Centre for Sustainable Development, University of Brasilia, Brasilia, Brazil.

Peter A. Singer, McLaughlin Rotman Centre for Global Health, University of Toronto & University Health Network, Toronto, Canada.

Joseph Wong, Department of Political Science, University of Toronto, Toronto, Canada.